A MESSAGE FROM MARTHA

A Message
from
Martha

THE EXTINCTION OF THE PASSENGER PIGEON

AND ITS RELEVANCE TODAY

MARK AVERY

B L O O M S B U R Y

LONDON · NEW DELHI · NEW YORK · SYDNEY

Published 2014 by Bloomsbury Publishing Plc,
50 Bedford Square, London WC1B 3DP

ISBN (print) 978-1-4729-0625-0
ISBN (ebook) 978-1-4729-0626-7

www.bloomsbury.com
www.bloomsburywildlife.com

Bloomsbury is a trademark of Bloomsbury Publishing Plc

A CIP catalogue record for this book is
available from the British Library

10 9 8 7 6 5 4 3 2 1

Contents

Preface

The owner of a motel in 29 Palms, California, near the Joshua Tree National Park, asked me why I was writing a book on an extinct bird when there are so many living birds to write about. I guess the answer is three-fold.

First, the Passenger Pigeon was the commonest bird in the world, and phenomenally abundant, and yet it has now been extinct for a century. The story of how it lived and how we drove it to extinction fascinates me, even though bits of the story are missing, and others need to be filled in by guesswork. Had you lived in the USA in 1800, or probably even in 1850, and had you been asked to name a bird that would be extinct in the wild by Easter 1900, then it is almost inconceivable that you would have chosen the Passenger Pigeon.

Second, for a nature conservationist like myself, extinction is generally deemed to be a 'bad thing'. We should use the centenary of the extinction of the Passenger Pigeon to examine how bad a thing it really is – and, if we believe it is a bad thing, then we should do more to prevent future extinctions. Only by examining case studies such as that of the Passenger Pigeon can we assess the impact of the loss of wild beauty from our world. Most of that loss is careless and unnecessary – and yet it has happened time and again, and it keeps on happening. Only by making an honest assessment of those impacts can we look forward to the future with a clearer gaze.

Third, the loss of the Passenger Pigeon was not an event separate from everything happening in the USA in the nineteenth century. This was the century of the American Civil War and the abolition of slavery, the century in which the West was won. The USA grew from 16 states in 1800 to 45 in 1900, and its population grew from five million to 76 million. The pace of change was rapid, and the destruction of natural resources, of which the Passenger Pigeon was a casualty, happened at an astonishing speed. In Europe, we gradually diminished our nature over centuries – but the American way was to rid the continent of vast areas of wilderness, and the species that lived there, within a generation. This book is a lot about the Passenger Pigeon, but also quite a lot about the USA in which the Passenger Pigeon lived and died, and died out.

I didn't say all that while checking in, of course. But if I had, I think my host might next have asked what made me, an English biologist, rather than an American historian, tackle the subject. That would be a fair question, and the only answer I can give is that it was the bird, the country and the recency that made me do it. The Passenger Pigeon had a way of life like no other on the planet, and we have lost its wonder forever. America is a fascinating mixture of bigotry and friendliness and grandeur and banality, and this was a very American extinction. And it all happened so recently. There were still Passenger Pigeons in the forests of Wisconsin and Michigan when each of my grandparents was born. Who wouldn't want to tell those stories?

In this book, I take a particular line on certain issues and I might as well explain some of them up front. I describe the Europeans who populated what is now the USA as European invaders, rather than as settlers or colonists. If, in the years after 1492, increasing numbers of Native Americans had landed in Europe and built cities or roamed the countryside trapping Foxes and Rabbits; if they had tricked us Europeans into treaties that set the rules of engagements between our two peoples and then systematically reneged on them to our great disadvantage; if they had subdued us with force and instigated policies within their own legislatures that required our removal from our ancestral lands – then we would describe them as invaders. They would be efficient and successful invaders, but invaders nonetheless.

I follow the handy American usage of referring to 'extinction' when I mean the global loss of a species and 'extirpation from [named place]' as the equivalent of local extinction. Moreover, although we often use the terms 'going extinct' or 'went extinct', you will not find them in this book because I do not believe that species go gentle into that bad night of extinction. They are pushed. And therefore I use the terms 'driven to extinction' and 'made extinct' quite a lot. That might irritate you a little as you read through this book, so I mention it now by way of explanation.

The Passenger Pigeon's main foods, which shaped its existence, were beech mast (beechnuts), oak acorns and chestnuts. I refer to these by their common names often, but collectively they are sometimes referred to as 'tree mast' in order to save space.

I am a scientist by training, trying to be a writer, after being a professional conservationist, and needing to be a bit of a historian to produce this book. Since it is a truth universally acknowledged that no mere man can multitask, this project was doomed from the start – but it's the best I could do.

Mark Avery, Northamptonshire

@markavery

March 2014

CHAPTER ONE

Introduction

The pigeon was a biological storm.

Aldo Leopold

On 1 September 1914, between midday and 1 pm, in the Cincinnati Zoo and Botanical Garden, Cincinnati, Ohio, a pigeon breathed her last, and with her died her species.

The pigeon was known as Martha, and the species was the Passenger Pigeon. Amongst all extinctions, this example remains unusual in two respects: the precision with which the timing is known, and the overwhelming abundance of the species just a few decades earlier – for, just a few decades before Martha died, the Passenger Pigeon was the commonest bird on Earth.

The last reliably recorded wild Passenger Pigeon had died a few years earlier, in 1900, in Pike County, Ohio. We can regard the record as reliable, as the bird was shot and the mounted skin can be seen in the Ohio Historical Society headquarters in Columbus. This Passenger Pigeon was also a female, and she was named 'Buttons', because the mounted specimen originally had buttons placed where her eyes had once been. Although a number of sightings of Passenger Pigeons were reported over the next few years none has been regarded as believable – and so the species lived on only in captivity, and from 1902 solely in Cincinnati Zoo, through the early years of the twentieth century.

The Passenger Pigeons in Cincinnati Zoo were all of captive origin, and they played out their lives without ever casting a shadow on the forests that were their natural home. A century earlier the Passenger Pigeon was so abundant that its flocks would block out the sun for hours at a time, as they passed by in numbers which are almost unimaginable for those of us who did not witness them, and which were almost incalculable even for those who did.

A century has passed since we lost the Passenger Pigeon, and I hope that the centenary might prompt us to consider what the loss of the commonest bird in the world tells us, if anything, about our own species' stewardship of the planet. What might be the message from Martha the Passenger Pigeon? That is the theme of this book, although in Chapter 6 I will introduce you to another Martha who fills in some more of the story.

This book is mainly about two species, the Passenger Pigeon (*Ectopistes migratorius*) and our own (*Homo sapiens*), although many others play their parts in the story. It was probably as late as 1850 that the global human population first exceeded that of the Passenger Pigeon, and only in the 1880s or 1890s that the human population of the USA exceeded that of the bird. It's a little like that in this book. At the beginning we are dealing very much with the Passenger Pigeon, but by the end we are focusing much more on ourselves.

Chapter 2 describes the basics of Passenger Pigeon biology, and also explores the ways in which this species was adapted to the forests of eastern North America.

How many Passenger Pigeons once existed on Earth is the subject of Chapter 3.

In Chapter 4 we go on a journey through the present-day United States, visiting places that were of relevance in the Passenger Pigeon story to gain greater insights into the bird and its fate.

Chapter 5 addresses the causes of the extinction of the Passenger Pigeon. The puzzle lies not so much in why the species declined, because there are so many reasons for that, but more in how we managed to finish it off so quickly, in a matter of a few short decades. There's quite a lot about our species in Chapter 5, because it was us who drove the Passenger Pigeon to extinction.

In Chapter 6 we go on another journey, but this time it is through time, to explore the events – ecological, social and political – which affected the Passenger Pigeon and many other species in the years running up to 1914. There is less about the Passenger Pigeon and more about us in this chapter, but also more about the Bison, the Carolina Parakeet, the Eskimo Curlew, the Rocky Mountain Locust and the Xerces Blue butterfly, as well as references to 'Cowboys and Indians', Jesse James and his gang, General George Armstrong Custer and a variety of presidents.

In Chapter 7 we address the question 'does extinction matter?' I have a strong view on this – and although describing the fate of the Passenger Pigeon has modulated that view a little, it is up to each of us to come to our own answer on that question.

Finally, Chapter 8 brings me, and English readers of this book, back to present-day conservation issues nearer to home, and in particular to the rapid decline of the Turtle Dove in English farmland around where I live. Will this be another Passenger Pigeon, and this time will we care, and will we save it? This is the real test. A hundred years on, are we listening? Will we hear the message from Martha and act upon it?

What of Martha herself? After death her body was sent to the Smithsonian Institution in Washington DC encased in a block of ice. She was measured, mounted and displayed for many years, but she is now hidden away from the public gaze.

CHAPTER TWO

The biology of an extinct bird

There are known knowns – these are things we know that we know.
There are known unknowns – that is to say, there are things that we
now know we don't know. But there are also unknown unknowns –
there are things we do not know we don't know.

Donald Rumsfeld

I have crossed the Atlantic Ocean ten times in my life, on five visits
to North America, the first of which was in the early 1980s. On
each outward journey I have flicked through the pages of a small
paperback book entitled *A Field Guide to the Birds East of the Rockies*
to help prepare myself for the new birds of an unfamiliar continent.
My copy, stained, well-thumbed, and much-loved, is the 1980
edition ('Completely new!' it says on the front cover) of Roger
Tory Peterson's classic field guide.

Peterson was born in Jamestown, New York, in the period
when, in today's conservation terms, the Passenger Pigeon was
'extinct in the wild' but when Martha, and others, lived on in
captivity. The first edition of his *Field Guide to the Birds* was published
in 1934 when Peterson was just 26. In the same year, Adolf Hitler
became Germany's head of state (he was already chancellor),
Bonnie Parker and Clyde Barrow were ambushed and killed in
Louisiana, a May dust storm removed huge amounts of topsoil
from the Great Plains, and F. Scott Fitzgerald published *Tender is
the Night*.

On page 181 of my copy are illustrated the familiar Rock Dove
(the ancestor of the feral pigeons that throng our cities) in the
bottom right, the White-winged Dove (which I have enjoyed
seeing in the deserts of Arizona) in the centre and, towards the top
left, the Mourning Dove, a common and familiar bird across much
of the USA and southern Canada. But circled in the very top

left-hand corner of the page is the head and neck of a Passenger Pigeon, labelled 'extinct'. It seems a quixotic gesture to keep the image of a long-extinct bird alive in the pages of a book designed to help us identify today's living birds. It is a gesture which I admire, and I wonder about the motivation of Peterson and his publisher. Did it have a deep motive? I like to think so – for, in what I am sure is a little-read note in the front of the book, Peterson wrote that 'the observation of birds leads inevitably to environmental awareness.'

Despite its presence in the Peterson *Field Guide*, we are a bit limited in how much we can observe the Passenger Pigeon these days – but a review of its biology can still lead to greater environmental awareness. Through examining the old records and thinking about the Passenger Pigeon in the light of modern biological knowledge, we can, even now, appreciate that this was not 'just another pigeon' – the details of its behaviour and ecology mark it out as being exceptional. When we drove the Passenger Pigeon to extinction we not only destroyed a unique species – every species is unique, after all – but also a uniquely interesting species.

This chapter describes and discusses the basic ecology of the Passenger Pigeon. However, before we learn about the bird itself we ought to dip into the life of the Passenger Pigeon's most prominent biographer, Arlie William Schorger, who was born in Republic, Ohio, in 1884 and died in 1972. From reading his obituaries in *The Auk* (the journal of the American Ornithologists' Union) and the *Passenger Pigeon* (the journal of the Wisconsin Society for Ornithology), it seems that Schorger was an interesting if not an easy man. 'Stubborn' is one of the words that come to mind when reading about Schorger. He stubbornly refused to wear a hearing aid in his later years despite poor hearing, and he stubbornly battled with editors to ensure that they changed little or nothing of his original carefully considered wording. Two colleagues smiled when they saw they were acknowledged for reading the draft of his book on the Passenger Pigeon, as neither could detect a single instance where he had changed anything in response to their comments.

Schorger spent most of his working life as a chemist. He specialised in waterproofing, and accumulated over 30 patents. But he was also a naturalist, and in 1951 he became a professor of wildlife management at the University of Wisconsin. A colleague, however, said that 'the content of his lectures was as excellent as his delivery was uninspiring', and he moved out of lecturing fairly quickly. It was during this time that he wrote *The Passenger Pigeon: its Natural History and Extinction*. The book was published in 1955, and for it he was awarded the American Ornithologists' Union's Brewster Medal in 1958.

One writer described Schorger as plain and unprepossessing, and others remarked that he rarely smiled. His prose has also been described as plain, and dispassionate, but maybe that added to its persuasiveness.

After reading all the scientific information about the Passenger Pigeon that he could lay his hands on, Schorger scanned the annals of the local newspapers for news items on the comings and goings of the species and the impacts of hunters. He spent his lunch hours, weekends and spare time for more than twenty years amassing information. This contrasts somewhat with the approach used by the author of this book, so I am glad to acknowledge a huge debt to Schorger. Not surprisingly, he was described as meticulous, diligent, thorough, untiring and precise.

Schorger's colleague at Wisconsin, Joseph J. Hickey, wrote of the book that 'there is no question about this being the final and definitive monograph on the species' – a view with which I would not argue. Indeed, as must any writer on the Passenger Pigeon, I have taken much of what Schorger wrote as almost gospel, and I have paused for a long time before departing from his views at all. But I hope that what this book can add now is a modern perspective on a long-gone species, and that I for my part can add a little of more modern biological thinking to the story than could Schorger, a chemist but a naturalist, around sixty years ago.

BIOLOGY OF THE BIRD

The Passenger Pigeon was a little larger than the Stock Dove of Europe, and quite a lot larger than the still-common Mourning Dove. Unlike the familiar pigeons and doves of the USA or Europe, the Passenger Pigeon showed noticable sexual dimorphism. The male had a beautiful grey-blue back and bright orange throat and chest, while the female, though only slightly smaller, was quite considerably duller in plumage. The male had bright orange eyes and coral legs and feet.

Both sexes had long broad wings and both had a notably long and graduated tail. Even now, even looking at a mounted specimen, this looks like a bird made for speed, with a powerful chest and a streamlined appearance. It deserves its alternative name of the 'Blue Meteor'.

John Muir (1838–1914), the Scot who moved to Wisconsin as a boy, and who did so much to establish the US National Park system, wrote as follows of the Passenger Pigeons he saw in those Midwest woods and fields in his teenage years:

The breast of the male is a fine rosy red, the lower part of the neck behind and along the sides changing from the red of the breast to gold, emerald green and rich crimson. The general color of the upper parts is grayish blue, the under parts white. The extreme length of the bird is about seventeen inches; the finely modeled slender tail about eight inches, and extent of wings twentyfour inches. The females are scarcely less beautiful. 'Oh, what bonnie, bonnie birds!' we exclaimed over the first that fell into our hands. 'Oh, what colors! Look at their breasts, bonnie as roses, and at their necks aglow wi' every color juist like the wonderfu' wood ducks. Oh, the bonnie, bonnie creatures, they beat a'! Where did they a' come fra, and where are they a' gan? It's awfu' like a sin to kill them!' To this some smug, practical old sinner would remark: 'Aye, it's a peety, as ye say, to kill the bonnie things, but they were made to be killed, and sent for us to eat as the quails were sent to God's chosen people, the Israelites, when they were starving in the desert ayont the Red Sea. And I must confess that meat was never put up in neater, handsomer-painted packages.'

Rather strangely, we do not have a very good estimate of how much Passenger Pigeons weighed – this is surprising for a bird that was often shot and trapped and ended up in pies. A weight of around 350 grams seems to be about right, with males being heavier and females lighter, and this weight seems in keeping with its body length of 41 centimetres for males and 35 centimetres for females (between the size of Europe's Wood Pigeon and Stock Dove).

The scientific name, *Ectopistes migratorius*, means 'wandering migrant', and we need to think of the Passenger Pigeon as a nomad, moving in great flocks to wherever its food was to be found. The main breeding range was approximately bordered on the east by the Atlantic Ocean and on the west by the Mississippi River, to the north by the Great Lakes and to the south by the southern borders of Kentucky, West Virginia, Pennsylvania and New York, but there are many records of nesting, including large colonies, away from this main breeding range. Outside of the breeding season the large flocks roamed widely to the north, the south and the west of the breeding areas. There was something of an autumn migration through August, September and October when flocks flooded down to the southeastern USA, but Passenger Pigeons were found widely (but thinly) across the Great Plains states, and the wanderings of some took them as far north as Hudson Bay.

This range reflects the distribution of the American Beech, Red Oak, White Oak and American Chestnut – the trees most important for the Passenger Pigeon. To the north, conifers began to predominate. To the west, grasslands took over, to the east was the ocean, and to the south, the Gulf of Mexico – and the ranges of all these trees ended in northern Florida. This was a bird of the eastern American deciduous forests.

The most important ecological factor determining the Passenger Pigeon's abundance and distribution was the production of mast by the American Beech, various oak species and American Chestnut. Beech and oak trees, the most important sources of food for Passenger Pigeons, vary greatly from year to year in their mast production, depending on the weather and other factors. Mature American Beech trees, the most favoured mast-producing species, only produce appreciable amounts of seed every 2–8 years, and these years of abundance are synchronised over hundreds of square kilometres. The life of a Passenger Pigeon was a constant search for areas rich in tree mast.

Wherever there had been a good beech or oak crop the previous autumn, that is where the pigeons would nest the next spring, as soon as the snow departed and the seed crop was exposed again to be exploited. Maybe the birds remembered the locations of good mast production from the preceding autumn, or maybe they reconnoitred each spring, or perhaps it was a mixture of the two.

Throughout the year, Passenger Pigeons fed in large flocks, scouring the forest floor for fallen tree mast. John Wheaton (1841–87) wrote of the feeding of Passenger Pigeons in an American Beech wood as follows:

> In the fall of 1859 ... I had an opportunity of observing a large flock while feeding. The flock ... present[ed] a front of over a quarter of a mile, with a depth of nearly a hundred yards. In a very few moments those in the rear, finding the ground stripped of mast, arose above the tree tops and alighted in front of the advance column. This movement soon became continuous and uniform, birds from the rear flying to the front so rapidly that the whole had the appearance of a rolling cylinder, having a diameter of about 50 yards, its interior filled with flying leaves and grass. The noise was deafening and the sight confusing to the mind.

A Passenger Pigeon swallowing a handful of acorns might store them for a while in its crop (where they aren't digested, just stored), and there are accounts of birds with their crops as large as oranges because they were so filled with food. The crops of well-fed birds that were shot sometimes burst on contact with the ground, and from them might spill 17 acorns or 28 beechnuts.

As well as the nuts of forests trees, other vegetable matter was readily consumed (as we might expect from a pigeon), including wild rice, every manner of berry such as cranberries in late summer, and the nuts and seeds of just about every deciduous tree (and some coniferous ones) in the Passenger Pigeon's range. They also supplemented their diet with insects, molluscs and worms.

Given the catholic diets of many pigeon species, it comes as no surprise that sometimes they switched their attention to cultivated crops. The Passenger Pigeon was a pest, at times, to those European invaders who were growing cereals. In Iowa, farmers delayed sowing wheat until after the pigeon migration so as to avoid losing the precious seed cast on the soil. Small boys were employed full-time to scare pigeons off seeded areas, and they had to get up early and work long hours to get their money, as a large flock of pigeons could do irreparable damage to any crop in just a few minutes. It is surely not a complete accident that the modern grain drill, which sets the seed into the soil out of the reach of surface-feeding birds, was invented in Horicon, Wisconsin, in 1860 – an area that was one of the last areas of abundance of the Passenger Pigeon. (Its inventor was Daniel Van Brunt, but he was not immortalised by having a band named after him like Jethro Tull, whose seed drill never really got off the ground – as it were.)

Passenger Pigeons would throw up the contents of their crops if they found more tempting food. This habit was said by farmers to be responsible for the spread of Pigeon Grass (or Green Foxtail) through arable areas of Michigan. And imagine the feelings of any farmer who saw flocks of pigeons regurgitating the rye seeds they had just taken from one of his fields so they could start on the wheat in another.

Birds do not have teeth, and therefore cannot chew, so in order to aid digestion they have muscular gizzards in which the food is ground down. After being stored in the crop, food passes into the stomach, where digestion starts, and then into the gizzard to be further broken down, and then it is passed back to the stomach. Many birds swallow small pieces of grit and store them in their gizzards to help with the grinding process. A handful of acorns does not look the most appetising of foods to me, nor the most easily digestible, and no doubt grit was important in facilitating the digestion of such tough food. Indeed, Passenger Pigeons used to visit islands in large rivers in huge numbers, both to drink the water and also to collect suitable grit.

Passenger Pigeons were greatly attracted to salt and would visit salt springs and salt marshes. There was some suggestion that when beech mast was their principal food they were even keener than usual to eat salt. This recalls the behaviour of some parakeets in South America, which eat clay to counter the effects of tannins in their diet.

The Passenger Pigeon only nested in areas of high mast production, but in such areas it nested in huge colonies, of millions of birds. Nesting occurred in the spring, from March through to June. The nests, high in the trees, were simple insubstantial collections of twigs. So insubstantial, in fact, that often one could easily see the sky, as well as the egg or chick, through the bottom of a nest above one's head. The male brought the materials for the nest and the female constructed it. Nest construction could be very rapid, with some nests completed within a single day.

A single white egg was laid in each nest, or perhaps sometimes two (we'll come back to this subject). The size of the Passenger Pigeon's egg was just about the size one would expect from a temperate-latitude pigeon of its size, around 38 by 26 millimetres.

Both sexes took turns in incubation. The female spent the night on the nest and the male roosted nearby. At dawn the males set out from the colony in large flocks to feed for several hours, returning later, and the females would depart, again in large flocks. Flocks of females would return in mid-afternoon to resume incubation until the following mid-morning, and the males would set out to feed before returning to roost in the colony at dusk. Depending on the time of day, during incubation, the colony was an all-male or all-female preserve, and the surrounding forests were populated by huge single-sex flocks of feeding birds.

After 13 days the egg would hatch, and hatching would be synchronised within each vast colony, so suddenly the woods were full of pigeon nests with young pigeons which needed feeding.

The young pigeons, known as squabs, were fed by both parents.

For the first few days, as in all pigeons, the squabs would be fed a protein-rich, fat-rich secretion from the adults' crops called 'pigeon milk' or 'crop milk'. Most birds feed their young on undigested or partially digested food items from the moment they hatch, but pigeons (and flamingos and penguins) produce this paste-like substance. The adult Passenger Pigeons ceased feeding for a few days before the egg hatched so that the crop milk was not contaminated with indigestible seeds or nuts. After only a few days the adults increasingly included adult food softened in the crop, and eventually, after about a week, the squabs were fed on regurgitated adult food.

The squabs got the food by sticking their beaks into their parents' throats as the crop milk was produced, and inevitably some was spilled. Shot adult Passenger Pigeons sometimes had curd-like strings of dried crop milk on their breasts – which informed their killers that the pigeons had squabs in their nests.

After 14 days there were no long goodbyes between parents and offspring, and this is another remarkable feature of the Passenger Pigeon life – the parents simply deserted the nest and left the young bird to fend for itself. Again, because of the synchrony of nesting, over a period of just a few days the colony would lose all its adults and there would be fat squabs everywhere, flightless and walking about on the ground after tumbling from their nests, but being able to fly after a period of another four days or so. The squabs were stuffed full of food by their parents before their departure, with their crops so full that observers described the distended crop as being equal in size to the rest of the young pigeon.

Schorger tells of Passenger Pigeon squabs sitting in their nests for a day after being abandoned and then fluttering to the ground, where within three or four days they could fly well enough to evade capture. Those sound like very dangerous days in the life of a Passenger Pigeon, but the immense size of the colonies meant that the local predators, we must assume, could only make a small impact on total pigeon numbers. Was this account, by William French, fanciful or accurate, I wonder?

When the young birds fluttered from the nests in large numbers they started at once and kept going ahead, in spite of the wild animals and hawks that killed many of them. If they came to a road they crossed it; a stream, they flew over; or they fell exhausted into the water and, flapping their wings, swam to the other shore and ran on into the night.

I would love to have seen this – and, if accurate, it describes behaviour that is quite unlike that of any other bird on Earth.

In the winter months, the pigeons generally headed south, but they moved to wherever there was food to be eaten. The main winter range spread down to northern Florida, Georgia, Louisiana and Texas. Notably, this was the part of the range where there was little if any snow cover and so the chestnuts, acorns and beech mast produced in summer were available to autumn and winter flocks.

Winter roosts were like summer nesting colonies – vast and communal. On the whole they were smaller (but still of a size that is difficult for us to imagine), and they were also more likely to be used regularly, off and on, over a period of several years. Peter Kalm described what appears to have been a winter roost:

In the spring of 1740, on the 11th, 12th, 15th, 16th, 17th, 18th and 22nd of March (old style), but more especially on the 11th, there came from the north an incredible multitude of these Pigeons to Pennsylvania and New Jersey. Their number, while in flight, extended 3 or 4 English miles in length, and more than one such mile in breadth, and they flew so closely together that the sky and the sun were obscured by them, the daylight becoming sensibly diminished by their shadow.

The big as well as the little trees in the woods, sometimes covering a distance of 7 English miles, became so filled with them that hardly a twig or a branch could be seen which they did not cover; on the thicker branches they had piled themselves up on one another's backs, quite about a yard high. When they alighted on the trees their weight was so heavy that not only big limbs and

branches of the size of a man's thigh were broken straight off, but
less firmly rooted trees broke down completely under the load.
The ground below the trees where they had spent the night was
entirely covered with their dung, which lay in great heaps. As soon
as they had devoured the acorns and other seeds which served them
as food and which generally lasted only for a day, they moved away
to another place.

Alexander Wilson (1766–1813), a Scot, born in Paisley, left
Scotland in his thirties and settled in Pennsylvania. His nine-
volume *American Ornithology* (1808–14) was the authoritative book
on American birds until Audubon published his work between
1827 and 1839. He wrote of a winter roost:

> It sometimes happens that, having consumed the whole produce of
> the beech trees, in an extensive district, they discover another, at
> the distance perhaps of sixty or eighty miles, to which they regularly
> repair every morning, and return as regularly in the course of
> the day, or in the evening, to their place of general rendezvous,
> or as it is usually called, the roosting place. These roosting places
> are always in the woods, and sometimes occupy a large extent of
> forest. When they have frequented one of these places for some
> time the appearance it exhibits is surprising. The ground is covered
> to the depth of several inches with their dung; all the tender grass
> and underwood destroyed; the surface strewed with large limbs of
> trees, broken down by the weight of the birds clustering one above
> another; and the trees themselves, for thousands of acres, killed as
> completely as if girdled with an ax. The marks of this desolation
> remain for many years on the spot; and numerous places could
> be pointed out, where, for several years after, scarcely a single
> vegetable made its appearance.

The Frenchman John James Audubon, who will make several more
appearances in this book, writes in his *Ornithological Biography* of
his sightings in Kentucky, not far from where Wilson had seen
Passenger Pigeons years earlier. He gives what appears to be the
most detailed account of a winter roost:

Let us now, kind reader, inspect their place of nightly rendezvous. One of these curious roosting places, on the banks of the Green River in Kentucky, I repeatedly visited. It was, as is always the case, in a portion of the forest where the trees were of great magnitude, and where there was little underwood. I rode through it upwards of forty miles, and, crossing it in different parts, found its average breadth to be rather more than three miles. My first view of it was about a fortnight subsequent to the period when they had made choice of it, and I arrived there nearly two hours before sunset. Few pigeons were then to be seen, but a great number of persons, with horses and wagons, guns and ammunition, had already established encampments on the borders.

And:

The dung lay several inches deep, covering the whole extent of the roosting place, like a bed of snow. Many trees two feet in diameter, I observed, were broken off at no great distance from the ground; and the branches of many of the largest and tallest had given way, as if the forest had been swept by a tornado. Everything proved to me that the number of birds resorting to this part of the forest must be immense beyond conception.

And:

The pigeons, arriving by thousands, alighted everywhere, one above another, until solid masses as large as hogsheads were formed on the branches all round. Here and there the perches gave way under the weight with a crash, and, falling to the ground destroyed hundreds of the birds beneath, forcing down the dense groups with which every stick was loaded. It was a scene of uproar and confusion.

And:

The pigeons were constantly coming, and it was past midnight before I perceived a decrease in the number of those that arrived.

The uproar continued the whole night; and as I was anxious to know to what distance the sound reached, I sent off a man, accustomed to perambulate the forest, who, returning two hours afterwards, informed me he had heard it distinctly when three miles distant from the spot. Toward the approach of day, the noise in some measure subsided, long before objects were distinguishable, the pigeons began to move off in a direction quite different from that in which they had arrived the evening before, and at sunrise all that were able to fly had disappeared. The howlings of the wolves now reached our ears, and the foxes, lynxes, cougars, bears, raccoons, opossums, and pole-cats were seen sneaking off, whilst eagles and hawks of different species, accompanied by a crowd of vultures, came to supplant them and enjoy their share of the spoil.

PIECING TOGETHER THE ECOLOGY

Those are the headline facts about this extinct bird. We know quite a lot about it, but certainly not everything we might wish. However, even with what we know, it is clear that the Passenger Pigeon was an exceptional bird: exceptionally gregarious, with exceptionally large colonies, exceptional nesting synchrony, and it was an exceptional wanderer too.

I am a biologist by training, so as I learned the facts of the biology of the Passenger Pigeon I was turning them over in my mind and trying to piece together the jigsaw of its ecology. The key to understanding the Passenger Pigeon, as with many other species, is to understand what it ate.

The Passenger Pigeon was a specialised species. It depended to a very large extent on eating tree mast, specifically beech mast, acorns and chestnuts. Its reliance on these trees influenced every aspect of its life.

The nomadic existence of the Passenger Pigeon was determined by the unpredictable abundance of tree mast – particularly beech mast. It could be a mast year in State A when it was a poor year in State B. Pigeons who hung around eking a living in State B when they could be gorging themselves in State A would leave fewer offspring than ones that evolved the nomadic habit of searching for the tree mast and exploiting its temporary abundance.

Let's first consider nomadism in the winter months to begin to understand the consequences for the species. After they had bred, vast flocks of millions, perhaps billions, of birds would head down to the southern states to spend the winter. We don't know whether the flocks remained intact or whether birds moved between them regularly, and we don't know how far they travelled through the winter. We do know that the flocks often made huge inroads into the mast lying on the forest floor. They were often compared to machines in their efficiency at removing every seed from an area where they had fed, sometimes to the detriment of free-range pigs inhabiting those same areas, and which were sometimes reported to have starved after Passenger Pigeons had cleared a locality of tree mast.

Let's explore the biology of this by imagining the flocks as combine harvesters and the wintering grounds as a big field of wheat ripe for the harvest. As a flock moved through the 'field' it would leave it practically bare of food until next year's crop was produced. This seems very efficient, but it also means that there would be a big advantage to flocks which remembered where they had been – there would be no point in revisiting areas that had already been depleted.

The field of wheat through which our avian combine harvester moved was not uniform. There were large bare patches where the trees had low mast productivity and other areas where food was abundant. What search strategy did the flocks use? Did they search systematically, quartering the 'field' and stopping where they spotted sufficiently abundant food to do some harvesting? How did they spot the areas which were rich in mast? Could they tell from looking down on the canopy whether there was food beneath it, or did they need to fly down into the woods to sample the abundance? I imagine they would, in their vast flocks, have been good at spotting signs of tree mast and diving down into the forests to feed on patches that were rich in their food. Sometimes Passenger Pigeon flocks flew low through the forests – was this when they were looking hard for mast? And sometimes they would fly high; was this when they had a decided destination 'in mind'?

Although the flocks of birds were huge, I assume there were many flocks, rather than all the pigeons in North America flying around their winter quarters in a single flock. Assuming that's correct, it is like letting several combine harvesters loose in the same field. How did that work? Did each flock search a particular part of the wintering range? Did the trajectories of the flocks crisscross and overlap so that one flock would sometimes be flying over an area that had been rich in mast but had already been harvested by another flock? Did they perhaps divide up the 'field' in some way – and if they did, how? I'd love to know the answers to these questions.

Birds have good spatial memory. At a small scale, this allows jays and tits, for example, to remember precisely where they have cached food so that they can retrieve it efficiently. At a larger scale they are good at finding their way on migration and remembering the details of feeding sites on the wintering grounds. I can easily imagine that Passenger Pigeons would be able to remember where they had already been during the winter and make good decisions on where to go next.

At the end of the winter the birds would move north to breed. But unlike most birds they weren't returning to the same place each year. In fact, wherever they had nested last year would probably be the worst place to nest this year, because it was unlikely to have had another good year for tree mast. Several authors have speculated that Passenger Pigeons might have remembered the locations of areas where tree mast was rich in the autumn and returned to nest in them in the spring. This seems perfectly possible.

But we can speculate that any Passenger Pigeon that lived to a good old age might well have covered a large part of the forests of eastern North America in its nomadic lifetime. An elderly Passenger Pigeon could have nested in half a dozen or more US states, and ranged even more widely outside the breeding area.

It doesn't seem to me to be stretching credulity too far to

imagine that older birds might behave as though they had a map of eastern North American forests in their heads which was annotated with recent nut crops. Did they think, or behave as though they thought, 'Michigan last summer so let's try Wisconsin this spring'? Such knowledge of the recent past could not be in the heads of young pigeons, of course, and we can imagine that 'follow older birds' would be a good rule of thumb for the youngsters.

How the leadership of a huge multimillion bird flock is organised is a matter beyond our comprehension. There must be some sort of leadership – if only because there are birds at the front and they are followed by birds behind them!

The Passenger Pigeons headed north, as the snow melted, to exploit the mast that had survived the winter. Small rodents, underneath the snow cover, would have made inroads into the mast abundance and as the snowmelt advanced other species – Wild Turkeys, squirrels, deer, and even bears – would also be feeding on the mast. This was an ever-decreasing resource throughout the pigeons' nesting season, so speed was of the essence.

There are many examples of Passenger Pigeons nesting and then failing because they were caught in late snowfalls. Schorger quotes the diary of Leroy Lyman from Pennsylvania, who wrote, 'About the middle of April, 1854, the pigeons commenced nesting near Coudersport, snow all gone. Soon after came high winds and for several days around zero weather, with snow several inches deep, breaking up the nesting and thousands of birds freezing to death. By May 20, the surviving birds were nesting again.' One reason for Passenger Pigeons risking early nesting would have been that their food supply was being depleted all the time. A leisurely approach to nesting would hardly be favoured under these ecological conditions – idling was a bad strategy.

There are several other aspects of the Passenger Pigeon's breeding strategy that I would interpret as having evolved for speed.

If the Passenger Pigeon really did lay just one egg then that's quite unusual for a pigeon living in temperate latitudes. Mourning

Doves, Stock Doves, Wood Pigeons, Turtle Doves and Collared Doves all lay a clutch of two. We'll come back to this subject a little later, because it is significant in another context, but the consequence of laying one egg instead of two would be to shorten the nesting attempt by a day or more, since birds almost always lay their eggs on successive days (or at longer intervals), and so it is certainly consistent with being a time-saving adaptation.

The 13-day incubation time certainly looks like an adaptation to the 'let's get on with it' way of life. It is about a day shorter than that of the Mourning Dove – a bird less than half the size of a Passenger Pigeon and one which you would expect to have a shorter incubation period. The open-nesting Turtle Dove, Collared Dove and Wood Pigeon have incubation periods of 15, 16 and 17 days respectively, whereas the hole-nesting Stock Dove idles through incubation for 22 days.

Squabs were fed for around 14 days – again, a very short period for a pigeon of this size. This also seems like an adaptation to shorten the time the adults needed to engage in a nesting attempt. The 14-day fledging period is as short as that of the Mourning Dove and much shorter than that of European pigeons, which range from 18–19 days for the smaller Turtle Dove to 33–34 days for the larger Wood Pigeon (and 35–37 days for the cave-nesting Rock Dove).

The Passenger Pigeon certainly got through a nesting attempt very quickly for a pigeon of its size, being able to build a nest, lay its egg, incubate its egg and then rear its squab all in 30 days. The necessity for such speed will certainly have been because it was exploiting an ever-decreasing food supply.

Passenger Pigeons nested in huge colonies, of millions of birds. Colonial nesting in birds is common, but certainly a minority activity. Only about one in eight of the world's 10,000 or so species of birds nest in colonies. Colonial nesting evolves when resources are clumped – usually food resources, but sometimes nesting resources too. Many seabirds are colonial breeders whose safe nesting locations (e.g. rat-free islands or predator-free cliffs) are restricted and whose marine prey are often clumped too. Species such as gannets and auks have traditional colonies which are relatively safe from predators and are within foraging range of good food supplies in most years – these birds' ability to forage over large distances reduces the impact of shifting prey populations on their breeding success. Terns are a little different. They nest together, seeking areas of low predator abundance, but defend their nests against predators – and that is one of the advantages of colonial nesting for them; but because terns' foraging ranges are quite small they shift between colonies at the drop of a hat, or more likely of a sprat. Many of the other colonial birds are waterbirds which nest close to wetlands predictably rich in food supplies (even if their richness might vary unpredictably from year to year).

Passenger Pigeons nested in trees and lived in a landscape dominated by trees – nest sites were hardly limited, it would seem. Schorger mentions nesting occurring in a wide range of tree species, and that colonies were often centred on river valleys but also sometimes on islands in lakes or rivers, or in marshes and swamps – but we can imagine that there were few physical barriers to the size of a Passenger Pigeon colony. And there is no evidence of any behaviours that would limit the numbers of pigeons nesting in a particular colony – birds would defend their own nest with a casual peck at any intruding bird, but apart from that it was open house.

Under these circumstances we can speculate on what limited or determined colony size. As described above for other birds, there would be two major factors which were important – reducing the impact of predators, and competition for food.

There is nothing to suggest that Passenger Pigeons ever actively defended their nesting colonies against predators, as do many colonial nesting species such as terns, waders and Fieldfares, but the mere presence of overwhelming numbers of nesting pigeons would swamp the impacts of local predators on their nests. Being with lots of other Passenger Pigeons was probably of advantage to each individual Passenger Pigeon, the weak and the strong, in terms of reducing their predation risk compared with nesting on their own.

The vast congregations of nesting pigeons provided a bonanza of food for local predators. Great numbers of eagles, Grey Wolves, foxes, hawks, mustelids, squirrels, rats and every other sort of predator that could eat full-grown or half-grown pigeons or their eggs would feast on the masses of birds in the air, in their nests or on the ground, but the impacts of such predators on the population of Passenger Pigeons must usually have been small because of the vastness of the pigeon numbers. Yes, the local hawks might have feasted on pigeons for a month when they nested in a location, but there is a limit to how many pigeons can be eaten by the local hawks, and hawks from a few territories away would not find the commuting a profitable way to feed their chicks. But this strategy depended for its success on there being an awful lot of Passenger Pigeons.

It seems that most colonies were long and thin rather than being approximate circles. This may be because they were often initiated in river valleys, perhaps because these areas were more sheltered. Schorger estimated the average area occupied by a colony as around 30 square miles, and that this would often lead to them being around 10 miles long and three miles wide (that's around 80 square kilometres: 16 × 5 km). Using estimates of nesting density from colonies, this would mean that a good working average of numbers for a Passenger Pigeon colony was 4,750,000 birds. This would be more than enough birds to swamp the impacts of local predators over a one-month period of nesting. Some of the largest colonies were estimated to hold more than 100 million birds, and although the absolute number of Passenger Pigeons killed by predators would presumably have been very large, the risk to an individual nesting bird was very, very small. The sheer numbers nesting in a colony diluted the impacts of predation on the colony as a whole.

There would, perhaps, have been things that an individual pair could do to minimise their risk even more. Nesting in the middle of the colony, away from the edges, would presumably often be safer. Perhaps nesting deep in the canopy would reduce the risk of predation by birds from above, and not being the lowest nesting bird might limit the risk from tree-climbing predators from below.

Being exposed far from others seems the worst position to be in, and it is interesting that there are accounts that suggest that the colonies had a sharp edge rather than petering out – with trees at the edge of the colony having 25 nests on one side of the tree and none on the exposed side. Nesting close (or sitting close, or flying close) to another Passenger Pigeon meant that any predator who saw you as its prey might decide that your close neighbour was a better bet. For a Passenger Pigeon to make the most of the arithmetic of the dilution effect it still wanted to be just one of the crowd – preferably one of the middle of the crowd.

Whereas it is easy to construct a line of reasoning (or at least a plausible story) to explain why colonial nesting could greatly reduce predation risk, it is more difficult to explain how living alongside millions of other birds, your closest competitors, would make feeding more profitable. And perhaps it didn't, but it was just too difficult to limit numbers, so every bird had to put up with the competition of others. Maybe, in anthropomorphic terms, every Passenger Pigeon in a vast colony wished that there were half as many, or maybe a tenth as many, other pigeons nesting at that site, but since none could influence that outcome they all had to put up with the hordes of competing birds.

The number of fellow Passenger Pigeons nesting around you would simply add to the incentive for speed in nesting. Not only was there no more food being produced, and not only were there other species eating it all the time, but you were surrounded by millions of direct competitors, too.

I have to believe that evolution favoured those Passenger Pigeons that nested in huge colonies, and so the balance of the advantages derived from nesting near a big supply of food and getting protection from predators had to outweigh the disadvantages of increased competition with fellow Passenger Pigeons.

However, just as in the winter we have to imagine several large flocks of Passenger Pigeons depleting the mast resource of the southern USA like combine harvesters in a vast field of wheat, so we have to imagine massive breeding colonies depleting the mast resources around those colonies. The bigger the colony, the faster the resources would be depleted, and so the larger the amount of resource needed within reach of the colony for breeding to be successful.

Imagine a single Passenger Pigeon pair (although the argument would apply to much larger numbers just as well) arriving at a well-established colony which was already 10 days into nesting. Should this pair join the colony? The arguments in favour are all to do with predator dilution. The pair would benefit from the protective presence of many other pigeons – but only at the beginning of the nesting attempt, because for the final 10 days of nesting our pair would be all alone, exposed to every passing predator, because the colony would have been deserted by the earlier-nesting birds. And our imaginary birds would also, in that 10-day period, be searching a landscape which had been already been stripped bare of adjacent and easily found tree mast by the earlier-nesting birds. The advantages of joining a colony fell off sharply as soon as the first birds began to nest, and that's what caused the Passenger Pigeon to evolve such a synchronous nesting biology. For late-arriving birds, the best option would often have been to move on and try to find another place to set up their own colony.

I think such a scenario must go some way towards explaining the amazing synchrony of the Passenger Pigeon colonies, in which all the eggs, often millions of them, were laid within just a very few days and therefore the young hatched synchronously, and the adults abandoned their offspring in a similarly short period. There was, it seems to me, a clear cost to being at the tail end of such a nesting – but what was the factor that caused the beginning of the nesting attempt to be so synchronised too?

We should imagine Passenger Pigeons at a nesting colony as being like competitors on their blocks at the start of a race, or eager shoppers waiting for the doors to open for the January sales. The starting gun was fired, or the doors flung open, by the snow melting from the trees (which would allow safe nesting) and from covering the tree mast on the ground (which would allow profitable feeding). There is little wonder that every Passenger Pigeon was ready to nest at about the same time – and then the race was on!

The size of colonies, and their spacing, would be the product of the individual decisions of lots of individual birds or pairs. However, to understand it better we could imagine being given the task of distributing the Passenger Pigeons around their breeding range. How would you go about it? You would need an accurate map of the distribution of food resources so that you could allocate birds in proportion to the resources. Where there were very large and very rich patches of tree mast you would place lots of birds (many colonies, or large colonies), where there were reasonable resources you would put fewer birds, and where the tree mast had failed (which would most likely account for large areas of the available habitat) you would distribute very few pigeons. If you were very good at allocating pigeons then it might be that each pair of pigeons ended up having equal access to resources whether it was in a huge colony or a small one – depending on your skill it would all even out in the end. And you would have to balance not just the access to resources but the dilution effect too. It would be a taxing task, but one which natural selection must have taken in its stride.

Just in passing, I will make the point that you might allocate some of your Passenger Pigeons very thinly across areas where mast was present but not in abundance. Maybe a proportion of the population would do better nesting outside of the colonies and forgoing the predator-diluting effect of nesting with others in favour of the reduced competition. Schorger supposed that the nesting success of such birds must have been very low – but that wasn't something he knew, only something he guessed.

Indeed, when it comes to knowing anything about the Passenger Pigeon nesting success then we are guessing. Most of what we know about the colonies, which isn't that much, came from the era when people were entering the colonies to collect squabs and kill adults, so they weren't necessarily the most attentive observers of the natural history of the birds. The observations weren't in place to tell us how often nesting failure happened in the vast colonies.

Nest sites were superabundant and there was no territorial behaviour to limit the number of breeders. It seems that there was nothing to limit the numbers of birds attempting to breed in any year. Under these circumstances we would expect that in some years, perhaps in years when overall the quantity of mast available happened to be low, and when by chance the preceding year had been very successful so that the pigeon population level was high, Passenger Pigeons could attempt to nest but would run out of food towards the end of the breeding season.

So, despite their almost unique nomadic lifestyle and their known ability to seek out areas rich in resources, might Passenger Pigeons sometimes have been caught out in poor resource years and experienced widespread nesting failure? It's possible. In fact I would say it's likely; but we don't know.

Of course, there would also have been very successful years, but despite the ability of the Passenger Pigeon to adapt to the boom-and-bust nature of its main food sources I would imagine that there were still boom-and-bust years of pigeon productivity, which would feed through to quite large fluctuations in population levels.

THE DEMOGRAPHIC CONUNDRUM

I'd love to know more about the Passenger Pigeon's ecology, but it seems to me that we have a fairly coherent view of it, considering that the species hasn't flown around the forests of eastern America for over a century. It was a gregarious nomad because its food was available in abundance each year but in different places each year. This also led to it being colonial, because it needed to exploit that abundance of food. Coloniality has advantages and disadvantages for the individual. Each bird benefits from the dilution effect, so its chance of being predated is greatly lessened by nesting with many others, but those many others are competitors for food too. And because the food was always being depleted, by other species and other Passenger Pigeons, this would have consequences for the nesting attempt – it was best to get it over and done with quickly before the ever-diminishing tree mast was too severely depleted.

The extreme synchrony of nesting within colonies was a product of the 'starting gun' of the snowmelt and the rapid fall-off in nesting success for any birds that started later than others in the colony. We can even imagine circumstances under which some Passenger Pigeons would nest outside of the enormous colonies.

It all hangs together quite well, it seems to me – though I'd love to be able to test these ideas in the field by studying real live Passenger Pigeons. However, there is one aspect of the species that still puzzles me a lot – let's call it the demographic conundrum – and this is where we finally get to grips with whether the Passenger Pigeon laid a clutch of one egg or two.

If the Passenger Pigeon laid a single egg (as Schorger believed, having sieved through the evidence) and nested but once each season (as Schorger also believed), then, for the species to maintain its numbers, the average annual survival would have had to have been so high that it strains credulity – and that's why it is a conundrum.

It's a commonplace observation, although rather interesting when you start to think about it, that many bird populations are relatively stable from year to year in the absence of human interference. This must mean that, in the long run, and quite often in the relatively short run too, their survival rates and productivity are in balance – losses through mortality are more or less balanced by the number of young birds that fledge from nests each year. If a million birds die each year but a million are born then the population remains at a constant level. It's a bit like Mr Micawber's comment about money: 'Annual income twenty pounds, annual expenditure nineteen [pounds] nineteen [shillings] and six [pence], result happiness. Annual income twenty pounds, annual expenditure twenty pounds ought and six, result misery.' For the birds, happiness would be an increasing population and misery a decreasing one.

Income and expenditure in this case are the productivity (number of young which fledge) and the death rate (the opposite of survivorship). The annual productivity of a pair of Passenger Pigeons would be the number of young they produce in the average season, which will be a factor of how many nesting attempts they make and how successful each of them is. But if you only lay one egg and only nest once, as Schorger suggested, then you will produce one fledged young a year – and that will only happen if all your eggs hatch and all hatched eggs lead to young fledging from the nest. Those young then need to survive until the next spring before they can enter the breeding population, and some of them will and some will not. And some adult birds will survive from one breeding season to the next and some will not.

Generally speaking, in fact just about always, in birds, adult survival is higher than first-year survival; adult birds are tougher, more experienced, wiser or have a combination of traits which make them a bit better at getting through life.

Nobody measured the survival of Passenger Pigeons in the wild, and so we can only make informed guesses about it. The European Stock Dove is about the same size as the Passenger Pigeon, and although we wouldn't have known much about its survival when the Passenger Pigeon was alive, nor really when Schorger wrote his book in the 1950s, we do now. The Stock Dove's vital statistics are roughly as follows. First-year survival (that's from being a chick in the nest to being an adult ready to breed) is around 40% (only four out of every ten young that leave the nest survive to join the breeding population the next spring), and adult survival is around 55%.

The larger Wood Pigeon (450 grams) survives a little better (52% first-year and 61% adult survival) as would be expected from a larger bird, whereas the smaller Turtle Dove (140 grams) survives a little less well (36% and 50% respectively). The Mourning Dove (130 grams) has a first-year survival of only around 30% and an adult survival of around 45%, as we would expect for a smaller but otherwise similar species.

Let's take the Passenger Pigeon as being fairly similar to the Stock Dove in survival rates. If we start the year with 100 Passenger Pigeons, by the same time next year only 55 of those birds will still be alive. However, they will have tried to nest, sometimes successfully, sometimes not, during the year and so will have produced some offspring. For there to be 45 offspring to fill the gap left by the deceased adults, at their lower first-year survival rate of 40%, there must have been 112.5 young produced by these 100 birds (50 pairs). This would be quite feasible for Wood Pigeons, Stock Doves and Mourning Doves, since all these species lay a clutch of two eggs, and nest several times each year, but it would be totally impossible for the Passenger Pigeon, which, we are told, laid a clutch of a single egg and only nested once a year.

Let us consider what a gap we have to fill. Our 50 pairs of Passenger Pigeons would, if Schorger is right, have laid 50 eggs. If we assume that they all hatched (which is very unlikely, given infertility, predation and harmful events such as late snowfalls) and all fledged (which is even more unlikely, what with predation and the food running out), then (given 40% first year survival like the Stock Dove) only 20 of them would enter the breeding population, and we need there to be 45 of them to balance the books. That is a chasm of a difference – instead of having 100 Passenger Pigeons the next year we only have 75.

Something doesn't add up here, and I've checked my sums, so it's not me! And that's why we have a conundrum.

There are three possible solutions. Maybe the Passenger Pigeon laid more eggs at each nesting attempt than we think was the case. Or maybe it nested more often each season than we think was the case. Or perhaps it had much higher survival than we would expect of a bird of its size living in temperate latitudes. Or it could have been some combination of the three – but something has to be quite different, and understanding that difference should help us to understand what factors may have led to the decline and extinction of this bird.

Did the Passenger Pigeon only lay one egg?

Schorger goes into quite some detail on the evidence, and he is pretty adamant in his view that they did, although he quotes various sources who say that at least sometimes a clutch consisted of two eggs.

Most captive birds only ever laid one egg, although there are reports of two, and indeed Audubon was a little miffed that captive birds he gave to various people on a visit to England only produced the one egg. In his *Ornithological Biography*, Audubon suggested that Passenger Pigeons laid two eggs, but Schorger points out that Audubon makes little of this and it doesn't appear to have been based on personal observations. Wilson, in contrast, was clear that the species laid just the one egg.

Even those accounts that suggest that two eggs were laid often stress that only one chick was ever produced, so for the purposes of this discussion that's the same as laying one egg. I'm inclined to take one potential chick fledged per nesting attempt as the normal maximum possible for the Passenger Pigeon. In addition, if the population were prone to big fluctuations in size, then perhaps in years of low population but abundant mast and acorns, two eggs were more often laid.

Nobody, to the best of my knowledge (and, more importantly, to the best of Schorger's knowledge), has ever suggested that the Passenger Pigeon laid three eggs, and it is clear that they did not always lay two. Therefore, even if some pairs, in some places, in some years did lay two eggs this doesn't help very much to balance the books of the demographic profit and loss account. Even if – and this is way too optimistic – the average clutch size was 1.5 eggs, it would lead to our 50 pairs laying 75 eggs, which (if they all hatched and fledged) could only lead to 30 additions to the breeding population in the next year, and we are looking for 45. So even by making some very unrealistic assumptions we have done little to close the gap.

Did the Passenger Pigeon nest successfully more than once a year?
The remarkably short time that the pair took to arrive at a nesting site, build a nest, incubate the (single) egg and then feed and desert the chick, a mere month in total, would leave plenty of time for early-nesting birds to move on and attempt to nest again – indeed, this contracted breeding timetable, with the single egg, the short incubation and the very short period of parental care, could be seen as adapted to a lifestyle of multiple breeding attempts in each

season. Many pigeon species have protracted nesting seasons during which there are multiple nestings, although few nest in colonies, and none in colonies quite like those of the Passenger Pigeon.

The nesting season in the northern part of the breeding range started in April, or perhaps late March, and that would leave open the possibility of nesting again in mid-May for successfully nesting birds, provided they could locate areas with abundant food supplies.

Schorger states that 'there was a general belief that the Passenger Pigeon nested several times a year. This is difficult to prove or disprove. Kalm stated that some people thought that they raised two broods each summer. Wilson was told that they bred three, and sometimes four times in the same season. Macauley goes the limit and has a hatching every month of the year.' Although Schorger admits to the possibility of repeat nesting by Passenger Pigeons, it is clear that he regards it as the exception rather than the rule.

Passenger Pigeons, like most birds, re-nested if they lost their first nesting attempt – which in their case was sometimes caused by heavy snow or other bad weather. In terms of annual productivity, this kind of re-nesting after a failure, often in the same colony or nearby (presumably because the area had been chosen for its abundant food, which was still present), is not the same as nesting successfully once and then moving to a new location and nesting successfully again. Indeed, these examples of sudden failure of nesting colonies due to weather events show how unlikely it would be that, on average, Passenger Pigeons would raise anything like as many as one fledged offspring from each nesting attempt – and this makes it even more difficult to make the sums on survival and productivity add up.

There are apparently no records of Passenger Pigeon colonies nesting successfully at a site and then simply nesting again in that site soon afterwards and producing a second crop of young. Not only has no-one observed and described such an event, but it seems ecologically unlikely that it could possibly happen – the resource depletion around a successful colony would surely have meant, since the tree mast in any given year was a non-replenishing resource, that a second nesting in the same location would involve huge distances of travel for the foraging parents.

If Passenger Pigeons nested more than once in the season then it seems that they upped sticks and searched for new places of mast abundance and set up new colonies. And that, in my view, is why they deserted their young in such an unusual manner and the adults all headed off – they were looking for new nesting opportunities.

The most likely pattern for such repeated nesting would surely be that the birds would nest as soon as they could in the southern part of the range where the snowmelt would occur first, or even in the wintering range, and then move north as the snows subsided. Schorger considered this possibility but found little evidence for a general northward movement of nesting dates in the records he examined, although those seemed rather sparse. There are records of colonies as far south as Oklahoma, but these records are not numerous, and the Oklahoma nesting commenced in mid-April – so it hardly seems that it was an early start, to be followed by those same birds moving hundreds of kilometres north to nest again (although it would be just possible).

Schorger did, however, document accounts from several observers of large numbers of young birds, birds of the year, arriving in nesting colonies that were in the early stages of nesting. Without knowing the details of the behaviour of individual birds, which we do not, this doesn't necessarily mean that individuals nested multiple times a year, but it is highly suggestive of that possibility.

These accounts came from a variety of sources. William Brewster (1851–1919, a co-founder of the American Ornithologists' Union) quotes S. S. Stevens of Cadillac, Michigan, a veteran pigeon netter of 'high reputation for veracity and carefulness of statement', and someone who had at least handled large numbers of Passenger Pigeons (dead and alive), as saying that pigeons 'breed during their absence in the South in the winter' because 'young birds in considerable numbers often accompany the early spring flights'. Native American writers also said that large numbers of young birds, less than a year old, arrived at a colony in Michigan about a week after the colony began nesting, and that similar things had been seen by others. These records of young of the year being seen in large numbers at colonies in Michigan argue strongly for earlier successful nesting – probably further south in the range.

Schorger's focus was on the records he could find in the local press of his adopted state of Wisconsin and nearby Michigan, and also his native Ohio. Although I admire (immensely) Schorger's (immense) compilation of the evidence on the Passenger Pigeon, I believe that his opinions were mostly formed by what he discovered in the accounts of the species on the northern edge of its range, and in accounts predominantly from the last 50 years of the birds' existence, and this focus may not have been best-suited to recognising, in the scraps of information that were available, any evidence for multiple nesting of the same birds in widely dispersed localities within the same year.

Schorger would also, in the early 1950s, have been ill-equipped to imagine the possibilities of the breadth of variation in avian breeding biology. Much of our knowledge of its complexity and range has come from studies dating from the 1970s and later. It is easier for us to be open to the possibility of multiple widely dispersed nesting attempts than it would have been for Schorger. Such multiple nesting attempts would still be unusual amongst birds, although not unparalleled.

The Mourning Dove lives a very different life from that of the Passenger Pigeon, nesting in scattered pairs rather than massive colonies. It is described as the 'champion of multiple brooding', being able to nest throughout the period between April and September (thus resembling the Collared Dove of Europe in the length of its breeding season). Mourning Doves are able to begin laying a new clutch (of two eggs) 3–6 days after successfully breeding, which suggests that, perhaps, Passenger Pigeons might have been capable of something similar.

The Quail, a European gamebird, is thought to nest in the south of its range in early spring and then move north hundreds of kilometres to make more nesting attempts. In some exceptional years Quail are common in the UK countryside, but generally speaking their northward wandering stops short of the UK. Although quails do not nest colonially they have a similar strategy to that which I am suggesting for the Passenger Pigeon, as this behaviour allows Quails to tap into the flushes of insects and grass seeds that make up its food.

And there is something else that Quail do which is remarkable, and I wonder whether it might have applied to Passenger Pigeons too. There are records of Quail apparently breeding in the north of their range, late in the season, which appear to be breeding attempts by the young of earlier broods of that same season. In other words, these birds are first nesting at the age of only several weeks rather than in the year following that in which they hatched. Might Passenger Pigeons have done the same? If they did, then this would be one explanation for young of the year arriving in nesting colonies later in the season – by then they were nesting birds too. This remains highly speculative, however, and we'll probably never know. To be fair, we do not even know for certain that Passenger Pigeons nested at age one as wild birds, but this is the case for other temperate-latitude pigeons, and if Passenger Pigeons did not breed until they were two years old then the demographic conundrum becomes an even greater puzzle.

I cannot believe that the Passenger Pigeon was limited to a single nesting attempt each breeding season – it's not very pigeon-like and it doesn't fit well with the speed with which the Passenger Pigeon raced through its breeding attempts. Most authors, following Schorger, have taken the line that the Passenger Pigeon nested but once each season – but I was pleased to find, after I had reached the conclusion myself that this looked highly unlikely, one strong dissenting voice. Enrique H. Bucher is an Argentinian ecologist who studied the Eared Dove of South America and published, in 1992, a paper on the extinction of the Passenger Pigeon. Bucher carried out similar simple sums to the ones presented here, and arrived at the same conclusion 20 years earlier – that the Passenger Pigeon must have nested multiple times a year. He based his sums on the survival of the Band-tailed Pigeon, whose adult survival is higher than the values I have used, but he still calculated that the Passenger Pigeon would have had to nest successfully between 1.23 times and twice a year (on average) to maintain its population (as it clearly had done, for thousands of years). The maths suggests that the average pair of Passenger Pigeons must have nested successfully something like five times every three years, for the species to have persisted. This has implications for the types of factor which might have caused the Passenger Pigeons' decline and eventual extinction.

The only way to escape the conclusion of multiple nesting attempts each year would be to assume that the Passenger Pigeon had surprisingly high adult survival rates.

Did the Passenger Pigeon have exceptional survival?
Since we don't know with complete certainty how many eggs were usually laid by the Passenger Pigeon, nor how many times during the season it nested, it would be very unrealistic to think that we might know its survival rate, as this takes a lot of work and a lot of time to discover even these days with extant birds. The approach I used above, of taking values that are similar to those of similar species living in similar places, seems reasonable, although it falls into the category of informed guesswork – which, although much better than un-informed or ill-informed guesswork, is still guesswork.

There are a couple of reasons, but only weak reasons, to imagine that the Passenger Pigeon did have high survival. The first is based on ecological principles, and the second brings us back to Martha.

Very few birds die of old age. Birds do age, but for most of their lives, and for most species, they are just as adept at doing things and just as chirpy aged eight years as they are aged two years. Most birds die from predation, starvation or disease. It seems likely that the Passenger Pigeon's ecology made it individually less vulnerable to predation, because of the dilution effect. Perhaps its nomadic lifestyle made it less vulnerable to starvation too, as it could move at any time of year to wherever food was abundant, or at least present, in ways that other species do not. But its lifestyle would surely have made it more vulnerable to infectious diseases, since it spent all its time in close proximity to others of its species.

But what of Martha? She lived her whole life in captivity, and, according to some accounts, that was a lifetime of 29 years – although others believe that the records are so hopelessly confused (and now unavailable to check since they were lost in a fire) that the figure is unreliable (and that several pigeons may have played the role of Martha over the years). Bob Ricklefs showed that there is a general relationship between how long a species lives in captivity and how long it survives in the wild, so a 29-year-old Martha would suggest that wild Passenger Pigeons were quite long-lived too. But it's not very strong evidence for anything really.

We really don't know much about the survival rate of the Passenger Pigeon, then, but we can calculate how high it would have to be to resolve the demographic conundrum. If the Passenger Pigeon really did lay one egg and nest just once each year then an adult survival of 75% and a first-year survival of 55% would just about allow the numbers to add up if all eggs hatched and all young fledged. In practice we know that there would be losses through the incubation and nestling period such that, for the population to remain stable with a single egg being laid by each pair during the nesting season, first-year survival would have to be around 70% and adult survival around 80%.

An adult survival rate of 80% would put the Passenger Pigeon in the same league as the Peregrine Falcon. This seems unlikely, as we would expect a prey species to have a lower survival rate than a predator, but who would say it's impossible when we have no real data on which to base our decision?

My ecological intuition leads me to think that the Passenger Pigeon was an unusual pigeon in that it normally laid a single egg (which was an adaptation to feeding on a diminishing food supply) but a usual temperate-latitude pigeon in that it made multiple nesting attempts each year. The unusual aspect of these nesting attempts was that they were usually made hundreds of kilometres apart from each other in order to exploit the rich patches of tree mast. I would be surprised if the Passenger Pigeon had an unusually high survival. That's my attempt to solve the demographic conundrum, but I wish we could go out and test it with data collected in the field.

THE PASSENGER PIGEON RESEARCH PROGRAMME, PART 1: TO FILL THE GAPS IN KNOWLEDGE

As a scientist I would love to be able to set up a research programme to fill in the gaps in our knowledge about this exceptional bird and to resolve the puzzles about its biology which it has left in our minds, long after it left our skies.

We would soon know how much the Passenger Pigeon weighed – you don't need a particularly taxing research programme to get that very basic piece of information.

Any remaining doubts about clutch size would soon be resolved; I wonder whether we would find that Schorger was right that it only ever lays a single egg. Personally I would expect that sometimes, and in some places, perhaps in the very best years for tree mast, at least some birds would lay a clutch of two. The more interesting aspects of Passenger Pigeon biology would take quite a lot of research effort to discover. We would need large grants from funding bodies to put into the field an army of PhD students and more experienced biologists for a period of many years to be able to estimate survival, and to study the movements of the birds.

Much of what we have learned about how long birds live and their movements has come from ringing (known in America as banding). This involves attaching numbered metal rings to birds' legs. The rings are hard-wearing but very light – they are usually described as adding, in relative terms, about the same weight to the ringed bird as does a wristwatch to a person. By recapturing ringed birds, or finding their corpses, we have built up a much clearer picture of how long many species of birds live, and of their movements around the globe. Both are aspects that we would want to illuminate for the Passenger Pigeon, but in this instance simple ringing would be a rather inadequate way to get all the answers.

Most birds are fairly site-faithful. The general pattern is that birds return fairly close to where they were hatched when they are of breeding age – this attraction to the place of birth is called philopatry. In most bird species, males are a bit more philopatric than females (perhaps because there is an advantage in setting up your territory, a predominantly male activity, near your relatives), and then, once it has established its home that's where you will find that bird throughout its life or, if it's a migratory species, that's where it will return each year. Most biologists studying birds can build up a local population where most of the birds are ringed, and therefore over a period of time their life histories begin to be known.

To ring it, you must first catch your bird, and some species are very difficult to catch (how would you go about catching a Golden Eagle, for example?). This sometimes means that the main source of birds to ring is the young birds in the nest, because they can't get away. We could organise expeditions to ring young Passenger Pigeons very easily, and we would target that period when the adults have left the colony and the squabs come down from the trees but are flightless or can only fly weakly. This would be an opportunity for us to ring thousands and thousands of known-age Passenger Pigeons. Added to this, there is a wealth of information from the pigeon-trappers of the nineteenth century which would allow our study to catch our target species throughout the year (and after checking them for rings release them) wherever in their range they occurred. If it were legal to shoot Passenger Pigeons in our present-day imaginary world, then we would seek the cooperation of hunters to return rings from shot Passenger Pigeons, and much information on movements and longevity could be built up like this.

Bird ringing was just beginning to come into existence at the time that the Passenger Pigeon was made extinct in the wild. The UK and US ringing, or banding, schemes were both set up in 1909 when Martha and two other captive birds in Cincinnati Zoo were the only Passenger Pigeons on the planet. We have learned a great deal in the years since then, and ringing has been central to our growing understanding of how birds live. However, wonderful though it has been, ringing is a fairly blunt research tool for studying any species. Imagine what you could discover about people by tattooing them and hoping to come across those same people again to read their tattoos, or relying on reports from undertakers. If the undertakers were very cooperative you would soon build up a picture of survival and longevity, and you would learn about gross movements from the distances between where you originally tattooed someone and the place of death. You'd discover some interesting facts, although it would take quite a long time – but it would only be skimming the surface of human life. And birds dying in the wild aren't very often encountered by the equivalent of undertakers, so the number of ringing recoveries of dead birds is tiny. We would want to know much more about the Passenger Pigeon – and technology is now coming to our aid.

One technique, now widely being used, is to put a small data-logger on a bird – only recently have these been made light enough to be practicable for many species. The clever technology records where the bird is at predetermined intervals (for example daily, or sometimes every few minutes). Through these devices we can learn much more about the movements of birds, the routes they take and the speeds at which they travel – all things that would greatly illuminate the nomadic ways of the Passenger Pigeon. And these devices are relatively cheap. At £70 each, our grant proposal to study the Passenger Pigeon might accommodate the purchase of 500 such devices. But there is a rather big snag. In order to keep the weight of the devices as low as possible they store data but do not transmit it – we would have to retrieve the data-loggers and download all that information. And the Passenger Pigeon would be one of the most difficult species to recapture because of its vast numbers (finding your bird with the data-logger would be akin to finding a needle in a haystack) and its nomadic ways – the very thing which we wished to study using the devices.

Such data-loggers are opening up the mysteries of seabird

feeding movements, which have been very difficult for land-based biologists to study. Easily recaptured species such as the hole-nesting Manx Shearwaters are ideal for this approach, but the Passenger Pigeon, where our few birds with their data-loggers attached would disappear into huge flocks and could be anywhere in the forests of the USA, is not the ideal species for such a study.

We would need even fancier technology for our study – satellite transmitters. Again, these devices have only recently been developed in versions that are light enough to fit onto birds the size of Passenger Pigeons. One of the studies that has captured the imagination of many has been the British Trust for Ornithology (BTO) study of Cuckoo migration. From ringing we knew a little of where Cuckoos spent the winter but a few dozen satellite tags are opening up the details of their travels to sub-Saharan Africa and where they spend the winter months. Using these devices, we could gather similar data for Passenger Pigeons.

We would know, for the first time, how far individual Passenger Pigeons travel in their lifetimes, how far they roam each winter, and where individuals nest in successive years. Satellite tagging would resolve the issue of whether, and how often, individual birds make multiple nesting attempts, and it would begin to resolve the demographic conundrum. But such information does not come cheaply – satellite tags and a share of satellite time come in at £3,000 each, so we could buy far fewer than we would be able to purchase data-loggers – perhaps the budget would stretch to just a dozen or so.

Would we find that birds moved between different flocks, or would we find that particular collections of birds tended to stay together? Would we find any evidence that the winter flocks tended to keep to particular parts of the wintering range, or would the trajectories of the winter flocks crisscross each other? Whatever we found out would be new – and no doubt it would pose more questions that we would want to answer.

We would want to relate the behaviour and movements of the Passenger Pigeons to the seed production of the oak and beech trees that provide their food resources, so we would probably need to recruit some botanists with tree expertise onto our team. How good are Passenger Pigeons at finding all the hotspots of mast production – do they miss some? Do Passenger Pigeons really behave as though they can calculate the best places to nest, taking into account food resources and colony size? How do the birds at any particular colony exploit the resources around them? How far do they travel to find food, and how efficiently and completely do they harvest the mast?

None of these questions were answerable when the Passenger Pigeon was still nesting and feeding, but then the state of biological knowledge was also much lower, and no-one would have thought of asking these types of question. The Passenger Pigeon was still a very common bird in 1859 when Charles Darwin's *On the Origin of Species by Means of Natural Selection, or the Preservation of Favoured Races in the Struggle for Life* was published. Schorger's book on the Passenger Pigeon was published in the year after David Lack's book *The Natural Regulation of Animal Numbers*, and Lack laid the foundation for a better understanding of how reproductive behaviour had evolved and what determines the levels of animal populations. But it was not until the late 1970s and 1980s that an explosion in what is called behavioural ecology added fine detail to our knowledge of how well birds have adapted to their environments through natural selection.

When the Passenger Pigeon was gracing the skies of North America, then, we neither had the technology nor the understanding of nature to ask sophisticated questions about its way of life. Now we are well armed to answer such questions – but we lack the Passenger Pigeon.

AN EXCEPTIONAL PIGEON AND AN EXCEPTIONAL BIRD

The Passenger Pigeon was a highly exceptional bird. It was exceptionally numerous (and quite how numerous is the subject for the next chapter) but it also had a very unusual way of life.

It depended, almost entirely, on tree mast for its food, which made it a highly specialised species. And the mast that it ate was the crop that the trees had produced in the previous year. It was, therefore, dependent on a food resource that was only renewed once a year. That's quite different from many colonial species who choose their nesting sites, such as wetlands, because they are rich in food and more food will be produced in them as the breeding season progresses. Not so with the Passenger Pigeon. When a pair of Passenger Pigeons chose to nest with millions of other Passenger Pigeons they 'knew' (or we could expect them to have evolved to behave as if they knew) that every item of tree mast within range of the colony was only there until something ate it – and the most likely and numerous competitors were all its fellow Passenger Pigeons.

The closest you get to this way of life in other birds is in the Red-billed Quelea, a sparrow-like bird from Africa, the Bronzewing Dove and Flock Pigeon of Australia, and the Eared Dove of South America. These are grassland species that are opportunistic breeders because their food supply is unpredictable, and they all nest in large (although not mind-bendingly vast) colonies.

The Red-billed Quelea is thought, now, in the absence of the Passenger Pigeon, to be the commonest bird in the world. Its population is estimated, with quite a lot of guesswork, to be in the order of 1.5 billion pairs. Studies by Peter Ward have pieced together the information about Red-billed Quelea breeding.

The Red-billed Quelea nests in colonies consisting of from hundreds to millions of birds, and each female lays 2–4 eggs per nesting attempt. The colonies are highly synchronised, as were those of the Passenger Pigeon, and each breeding attempt is rapid, as was the Passenger Pigeon's, with incubation of around 10 days and a fledging period of around 16 days – like the Passenger Pigeon, they can produce their young in around a month, and again like the Passenger Pigeon, they abandon their young. Other similarities are that for the Red-billed Quelea nest sites are not limiting (any old bush or tree will do), in the largest colonies predators are probably swamped and diluted, and there is little fidelity to breeding sites from year to year.

Red-billed Queleas nest several times during the season, moving from place to place where their food (grass seeds) is abundant. Towards the end of one nesting attempt, nesting females were shown to have three egg yolks developing and were thought to be biologically able to nest again in a matter of days. This seems to be what Red-billed Queleas do – Michael Jaeger marked nesting birds in colonies in Ethiopia with paint in June and found many of these same birds nesting in colonies 800–1100 kilometres further north 100 days later. Red-billed Queleas in the arid areas of Africa have a nomadic breeding system which follows the rains and the pulses of food supplies that the seasonal rains create in dry areas. This is a somewhat similar system to the nomadic multiple nesting that I suggest for the Passenger Pigeon in the forests of northeast America, except that Red-billed Queleas are exploiting new pulses of abundant food created by the seasonal rains, whereas Passenger Pigeons were exploiting a depleting source of food made available by the melting snows.

The Red-billed Quelea, again like the Passenger Pigeon, has a liking for human crops, such as sorghum in its case, and therefore its vast numbers are regarded as a scourge rather than a boon by local farmers. It doesn't seem as though the Red-billed Quelea tastes particularly nice, either. And so the Red-billed Quelea is not hunted for its flesh, but it is destroyed because of its impact on farming production. Despite being sprayed with pesticides, and despite its colonies being set on fire and dynamited, the Red-billed Quelea does not appear to be moving rapidly towards extinction, even though people have tried hard to reduce its numbers. It makes you wonder how the Passenger Pigeon was lost so quickly from the Earth when no-one was aiming for its extinction, and when it was several times more numerous than the Red-billed Quelea at its highest point, and it lived in a time of less technological sophistication when it came to killing birds – but that is the subject of Chapter 5.

Was the Passenger Pigeon the commonest bird the world has ever known? That is the subject we tackle in the next chapter.

CHAPTER THREE

Lost abundance

The wisdom of crowds.
James Surowiecki

At 1 pm on 1 September 1914 there was no longer a living Passenger Pigeon on Earth. Less than a century before, there had been individual flocks comprising billions of Marthas. How abundant was the Passenger Pigeon?

Schorger plumped for a figure of 3–5 billion Passenger Pigeons as his estimate of the population before the European invasion of the USA. How reasonable a figure is that?

Accounts of flocks and breeding colonies are our starting place for an estimate of the Passenger Pigeon population, but right from the start we must realise that although these will amaze us with the sheer numbers of birds that were present, they will not, of themselves, reveal the total population levels.

Imagine a Passenger Pigeon trying to estimate the human population of the USA. Stories of huge numbers of people in New York City or Los Angeles might impress other pigeons, but if that were all they had to go on then they would find it difficult to reach an estimate of the total human population of the country. That is the position in which we shall find ourselves after examining the historical accounts, but they will form the basis for a partial unravelling of another conundrum – the Passenger Pigeon's population size.

EARLY ACCOUNTS

The Passenger Pigeon was noticeably numerous when the early European explorers encountered the New World. The pigeons were common or abundant in some of the areas first explored, notably Prince Edward Island, Quebec and Maine.

The first records of Passenger Pigeon from Europeans were of birds seen in 1534 by Jacques Cartier at Cape Orleans, now Cape Kildare, on Prince Edward Island – near the Provincial Park which now bears his name. Cartier, from St Malo, Brittany, reported 'infinite' numbers of 'wood pigeons' at the site, and again when he travelled down the St Lawrence River. Seventy-odd years later, in the summer of 1605, Samuel de Champlain saw countless numbers of pigeons at Kennebunkport, Maine. As the European invaders settled in New England they reported back on the abundance of wildlife, and the Passenger Pigeon was often mentioned. Sometimes they were compared with the Stock Dove, and rather more often with the Turtle Dove of Europe, but often they were described just as pigeons or wild pigeons. Further early reports come from Virginia, Manhattan Island, New York (when it was still New Amsterdam), Delaware and Massachusetts.

The pigeons were 'very plentiful in the woods', of 'countless numbers' and an 'infinite multitude'. They were regarded as pests that ate the European invaders' crops, but also as a boon, being at times so numerous that they were easy to catch and add to the pot, and of 'an excellent taste'.

The nestings

The breeding colonies, or 'nestings', were described at length by many observers, and it is worth transcribing some of these accounts in detail, to convey the full flavour.

An early account of a colony was that of Peter Kalm (1716–79), a Swede who studied under Carl Linnaeus and settled in America:

> When toward the end of June, 1749 ... I had left the English Colonies, and set out for Canada through the wilderness which separates the English and French Colonies from each other, and which to a great extent consists of thick and lofty forests, I had an opportunity of seeing these Pigeons in countless numbers. Their young had at this time left their nests, and their great numbers darkened the sky when they occasionally arose *en masse* from the trees into the air. In some places the trees were full of their

nests. The Frenchmen whom we met in this place had shot a great number of them, and of this they gave us a goodly share. These Pigeons kept up a noisy murmuring and cooing sound all night, during which time the trees were full of them, and it was difficult to obtain peaceful sleep on account of their continuous noise. In this wilderness we could hear in the night time, during the calmest weather, big trees collapsing in the forests, which during the silence of the night caused tremendous reports: this might in all probability be ascribed to the Pigeons, which according to their custom had loaded a tree down with their numbers to such an extent that it broke down.

James Fenimore Cooper (1789–1851) wrote in his novel *The Chainbearer*, published 1845, of a nesting in central New York State:

I scarce know how to describe the remarkable scene. As we drew near to the summit of the hill, pigeons began to be seen fluttering among the branches over our heads, as individuals are met along the roads that lead into the suburbs of a large town. We had probably seen a thousand birds glancing around among the trees, before we came in view of the roost itself. The numbers increased as we drew nearer, and presently the forest was alive with them.

The fluttering was incessant, and often startling as we passed ahead, our march producing a movement in the living crowd, that really became confounding. Every tree was literally covered with nests, many having at least a thousand of these frail tenements on their branches, and shaded by the leaves. They often touched each other, a wonderful degree of order prevailing among the hundreds of thousands of families that were here assembled.

The place had the odor of a fowl-house, and squabs just fledged sufficiently to trust themselves in short flights, were fluttering around us in all directions, in tens of thousands. To these were to be added the parents of the young race endeavoring to protect them and guide them in a way to escape harm. Although the birds rose as we approached, and the woods just around us seemed fairly alive with pigeons, our presence produced no general commotion; every one of the feathered throng appearing to be so much occupied with its own concerns, as to take little heed of the visit of a party of strangers, though of a race usually so formidable to their own.

The masses moved before us precisely as a crowd of human beings yields to a pressure or a danger on any given point; the vacuum created by its passage filling in its rear as the water of the ocean flows into the track of the keel.

The effect on most of us was confounding, and I can only compare the sensation produced on myself by the extraordinary tumult to that a man experiences at finding himself suddenly placed in the midst of an excited throng of human beings. The unnatural disregard of our persons manifested by the birds greatly heightened the effect, and caused me to feel as if some unearthly influence reigned in the place. It was strange, indeed, to be in a mob of the feathered race, that scarce exhibited a consciousness of one's presence. The pigeons seemed a world of themselves, and too much occupied with their own concerns to take heed of matters that lay beyond them.

Not one of our party spoke for several minutes. Astonishment seemed to hold us all tongue-tied, and we moved slowly forward into the fluttering throng, silent, absorbed, and full of admiration of the works of the Creator. It was not easy to hear each others' voices when we did speak, the incessant fluttering of wings filling the air. Nor were the birds silent in other respects.

The pigeon is not a noisy creature, but a million crowded together on the summit of one hill, occupying a space of less than a mile square, did not leave the forest in its ordinary impressive stillness.

Alexander Wilson, the father of American ornithology, wrote:

The breeding place differs from the former [i.e. winter roosts] in its greater extent. In the western countries above mentioned, these are generally in beech woods, and often extend, in nearly a straight line across the country for a great way. Not far from Shelbyville, in the State of Kentucky, about five years ago, there was one of these breeding places, which stretched through the woods in nearly a north and south direction; was several miles in breadth, and was said to be upwards of forty miles in extent! In this tract almost every tree was furnished with nests, wherever the branches could accommodate them. The pigeons made their first appearance there about the 10th of April, and left it altogether, with their young, before the 29th of May.

As soon as the young were fully grown, and before they left the nests, numerous parties of the inhabitants from all parts of the adjacent country came with wagons, axes, beds, cooking utensils, many of them accompanied by the greater part of their families, and encamped for several days at this immense nursery. Several of them informed me that the noise in the woods was so great as to terrify their horses, and that it was difficult for one person to hear another speak without bawling in his ear. The ground was strewed with broken limbs of trees, eggs, and young squab pigeons, which had been precipitated from above, and on which herds of hogs were fattening. Hawks, buzzards, and eagles were sailing about in great numbers, and seizing the squabs from their nests at pleasure; while from twenty feet upwards to the tops of the trees the view through the woods presented a perpetual tumult of crowding and fluttering multitudes of pigeons, their wings roaring like thunder, mingled with the frequent crash of falling timber; for now the ax-men were at work cutting down those trees that seemed to be most crowded with nests, and contrived to fell them in such a manner that, in their descent, they might bring down several others; by which means the falling of one large tree sometimes produced two hundred squabs, little inferior in size to the old ones, and almost one mass of fat. On some single trees upwards of one hundred nests were found, each containing one young only; a circumstance in the history of this bird not generally known to naturalists. It was dangerous to walk under these flying and fluttering millions, from the frequent fall of large branches, broken down by the weight of the multitudes above, and which, in their descent, often destroyed numbers of the birds themselves; while the clothes of those engaged in traversing the woods were completely covered with the excrements of the pigeons.

Charles Bendire, in the *Life Histories of American Birds*, described some observations made much later in the nineteenth century by the pigeon netter S. S. Stevens of Cadillac, Michigan:

> The largest nesting he ever visited was in 1876 or 1877. It began near Petoskey, and extended northeast past Crooked Lake for 28 miles, averaging 3 or 4 miles wide. The birds arrived in two separate bodies, one directly from the south by land, the other following the east coast of Wisconsin, and crossing at Manitou Island. He saw the latter body come in from the lake at about 3 o'clock in the afternoon. It was a compact mass of pigeons, at least 5 miles long by 1 mile wide. The birds began building when the snow was 12 inches deep in the woods, although the fields were bare at the time. So rapidly did the colony extend its boundaries that it soon passed literally over and around the place where he was netting, although when he began, this point was several miles from the nearest nest. Nestings usually start in deciduous woods, but during their progress the pigeons do not skip any kind of trees they encounter. The Petoskey nesting extended 8 miles through hardwood timber, then crossed a river bottom wooded with arborvitæ, and thence stretched through white pine woods about 20 miles. For the entire distance of 28 miles every tree of any size had more or less nests, and many trees were filled with them. None were lower than about 15 feet above the ground.

The density of nests could be enormous. Perhaps we can dismiss James Fenimore Cooper's description of 'at least a thousand of these frail tenements' as a novelist's exaggeration, but nonetheless accounts exist of 500 nests in a single tree. Schorger didn't believe this figure, but he did believe accounts of Native Americans taking around 200 young pigeons from particular trees, and he seemed to believe the account of 317 nests being found in a single Eastern Hemlock tree. There are many accounts of 100 nests from a single tree.

A nesting in Wisconsin in 1871 is generally regarded as the largest ever reasonably well documented. This colony had two arms stretching northwards, one to the east and the other to the west, from near the town of Kilbourn City. The eastern arm had a length of 50 miles and a width of eight miles, while the western arm was 75 miles long and around six miles wide. In all, then, this colony covered 850 square miles (slightly larger than the English county of Herefordshire), although there would have been gaps, holes and variations in the density of nesting birds through this area. Schorger also mentions that this was the 'main' nesting colony and that others existed nearby containing millions more birds. Perhaps we should regard Schorger's main colony as a cluster of colonies, maybe of 6–10 colonies spread over a large area but closer together than was sometimes the case. Such a dense gathering of birds, spread over such a large area, would be a challenge to count, and no-one attempted to do so. Given that we know that some individual trees would hold more than 100 nests, we can imagine that the number of birds was just vast.

But let's not just imagine – let's do some sums. 850 square miles is about 2200 square kilometres or 220,000 hectares. Let us say there are 20 trees per hectare – that is 4.4 million trees. If each tree has an average of 20 nests per tree that would be 88 million nests occupied by 88 million pairs of birds. That's 176 million adult Passenger Pigeons.

This can hardly be called an estimate – more like an amusement – and the final figure depends critically on the area covered by the cluster of colonies as a whole (the 850 square miles – which it is difficult to check, more than 140 years later), and within that general area on the tree density and the average number of nests per tree. It is surprisingly difficult to find estimates of tree density in old-growth forests, but those that exist suggest that there might well be at least 20 quite large trees per hectare (trees with a diameter at chest height of more than 40 centimetres – that is a pretty decent tree) and many smaller ones that would still be able to hold many pigeon nests. And then there is the difficulty of assessing how many nests one should allow per tree, taking account both of occupied and unoccupied areas of forest and of the variability in nests per occupied tree. It is more of a parlour game than a scientific estimate.

I was gratified to find that Schorger played the game too. His estimate of the number of Passenger Pigeons at this same colony was 136 million, which seems remarkably close to my own guess. We used the same area for the colony (of course! – I used the one from his book) but his estimate of trees per hectare was about three times that which I used and his estimate of nests per tree was a quarter of my guess, and so his estimate is around three-quarters of mine.

If we take Schorger's conservative estimate of a Passenger Pigeon population before the European invasion of three billion birds, and a rough estimate of 1,000 birds per occupied hectare of land in a colony, then we could fit our Passenger Pigeons into 30,000 square kilometres – an enormous area. If we allow the breeding range of the Passenger Pigeon to stretch as far west as Missouri, Illinois and Wisconsin and as far south as Missouri, Kentucky and Virginia, and then throw in a bit of southern Canada too (we know that breeding occurred outside this area, but probably not in very large numbers or with great frequency), then the land available amounted to around two million square kilometres. In any one year, therefore, the colonies would occupy 1.5% of the total breeding range. And if the pigeons nested twice each year, and used different locations for each nesting attempt, then the total used in each year would be around 3%.

The large flocks

It seems fitting to quote first from a Native American account of a flock of Passenger Pigeons. Simon Pokagon (1830–99) was a Potawatomi from southwest Michigan who campaigned for Native American rights and wrote many books and articles. Of the Passenger Pigeon, he wrote:

> When a young man I have stood for hours admiring the movements of these birds. I have seen them fly in unbroken lines from the horizon, one line succeeding another from morning until night,

moving their unbroken columns like an army of trained soldiers pushing to the front, while detached bodies of these birds appeared in different parts of the heavens, pressing forward in haste like raw recruits preparing for battle. At other times I have seen them move in one unbroken column for hours across the sky, like some great river, ever varying in hue; and as the mighty stream, sweeping on at sixty miles an hour, reached some deep valley, it would pour its living mass headlong down hundreds of feet, sounding as though a whirlwind was abroad in the land. I have stood by the grandest waterfall of America and regarded the descending torrents in wonder and astonishment, yet never have my astonishment, wonder, and admiration been so stirred as when I have witnessed these birds drop from their course like meteors from heaven.

And:

About the middle of May, 1850, while in the fur trade, I was camping on the head waters of the Manistee River in Michigan. One morning on leaving my wigwam I was startled by hearing a gurgling, rumbling sound, as though an army of horses laden with sleigh bells was advancing through the deep forests towards me. As I listened more intently I concluded that instead of the tramping of horses it was distant thunder; and yet the morning was clear, calm and beautiful. Nearer and nearer came the strange commingling sounds of sleigh bells, mixed with the rumbling of an approaching storm. While I gazed in wonder and astonishment, I beheld moving toward me in an unbroken front millions of pigeons, the first I had seen that season. They passed like a cloud through the branches of the high trees, through the underbrush and over the ground, apparently overturning every leaf.

Some years earlier, Alexander Wilson described observations made in the early 1800s:

I had left the public road to visit the remains of the breeding place near Shelbyville, and was traversing the woods with my gun, on my way to Frankfort, when, about one o'clock, the pigeons, which I had observed flying the greater part of the morning northerly, began to return in such immense numbers as I never before had witnessed. Coming to an opening by the side of a creek called the Benson, where I had a more uninterrupted view, I was astonished at their appearance. They were flying with great steadiness and rapidity at a height beyond gunshot in several strata deep, and so close together that could shot have reached them one discharge could not have failed of bringing down several individuals. From right to left, far as the eye could reach, the breadth of this vast procession extended, seeming everywhere equally crowded. Curious to determine how long this appearance would continue, I took out my watch to note the time, and sat down to observe them. It was then half-past one. I sat for more than an hour, but, instead of a diminution of this prodigious procession, it seemed rather to increase both in numbers and rapidity, and, anxious to reach Frankfort before night, I rose and went on. About four o'clock in the afternoon I crossed the Kentucky River at the town of Frankfort, at which time the living torrent above my head seemed as numerous and as extensive as ever. Long after this I observed them in large bodies that continued to pass for six or eight minutes, and these again were followed by other detached bodies, all moving in the same southeast direction, till after six in the evening.

And a little later:

Let us first attempt to calculate the numbers of that above mentioned, as seen in passing between Frankfort and the Indiana territory. If we suppose this column to have been one mile in breadth (and I believe it to have been much more), and that it moved at the rate of one mile in a minute, four hours, the time it continued passing, would make its whole length two hundred and forty miles. Again, supposing that each square yard of this moving body comprehended three pigeons, the square yards in the whole space, multiplied by three, would give two thousand two hundred and thirty millions, two hundred and seventy-two thousand pigeons – an almost inconceivable multitude, and yet probably far below the actual amount.

Wilson also wrote:

The appearance of large detached bodies of them in the air and the various evolutions they display are strikingly picturesque and interesting. In descending the Ohio by myself in the month of February I often rested on my oars to contemplate their aerial manoeuvres. A column, eight or ten miles in length, would appear from Kentucky, high in air, steering across to Indiana. The leaders of this great body would sometimes gradually vary their course until it formed a large bend of more than a mile in diameter, those behind tracing the exact route of their predecessors. This would continue sometimes long after both extremities were beyond the reach of sight, so that the whole, with its glittery undulations, marked a space on the face of the heavens resembling the windings of a vast and majestic river. When this bend became very great the birds, as if sensible of the unnecessary circuitous course they were taking, suddenly changed their direction, so that what was in column before, became an immense front, straightening all its indentures, until it swept the heavens in one vast and infinitely extended line. Other lesser bodies also united with each other as they happened to approach with such ease and elegance of evolution, forming new figures, and varying these as they united or separated, that I never was tired of contemplating them. Sometimes a hawk would make a sweep on a particular part of the column from a great height, when, almost as quick as lightning, that part shot downwards out of the common track, but soon rising again, continued advancing at the same height as before. This inflection was continued by those behind, who, on arriving at this point, dived down, almost perpendicularly, to a great depth, and rising, followed the exact path of those that went before. As these vast bodies passed over the river near me, the surface of the water, which was before smooth as glass, appeared marked with innumerable dimples, occasioned by the dropping of their dung, resembling the commencement of a shower of large drops of rain or hail.

Happening to go ashore one charming afternoon, to purchase some milk at a house that stood near the river, and while talking with the people within doors, I was suddenly struck with astonishment at a loud rushing roar, succeeded by instant darkness, which, on the first moment, I took for a tornado about to overwhelm the house and everything around in destruction. The people, observing my surprise, coolly said: 'It is only the pigeons'; and on running out I beheld a flock, thirty or forty yards in width, sweeping along very low between the house and the mountain, or height, that formed the second bank of the river. These continued passing for more than a quarter of an hour, and at length varied their bearing so as to pass over the mountain, behind which they disappeared before the rear came up.

In his *Ornithological Biography*, Audubon wrote of his sightings in Kentucky, not far from where Wilson had seen and attempted to count Passenger Pigeons a few years earlier:

The multitudes of wild pigeons in our woods are astonishing. Indeed, after having viewed them so often, and under so many circumstances, I even now feel inclined to pause, and assure myself that what I am going to relate is fact. Yet I have seen it all, and that, too, in the company of persons who, like myself, were struck with amazement.

In the autumn of 1813, I left my house at Henderson, on the banks of the Ohio, on my way to Louisville. In passing over the Barrens a few miles beyond Hardensburgh, I observed the pigeons flying from northeast to southwest, in greater numbers than I thought I had ever seen them before, and feeling an inclination to count the flocks that might pass within the reach of my eye in one hour, I dismounted, seated myself on an eminence, and began to mark with my pencil, making a dot for every flock that passed. In a short time, finding the task which I had undertaken impracticable, as the birds poured in in countless multitudes, I rose, and counting the dots then put down, found that one hundred and sixty-three had been made in twenty-one minutes. I traveled on, and still met more the farther I proceeded. The air was literally filled with pigeons; the light of noonday was obscured as by an eclipse; the dung fell in spots, not unlike melting flakes of snow; and the continued buzz of wings had a tendency to lull my senses to repose.

Whilst waiting for dinner at Young's Inn, at the confluence of Salt River with the Ohio, I saw, at my leisure, immense legions still going by, with a front reaching far beyond the Ohio on the west, and the beechwood forests directly on the east of me. Not a single bird alighted; for not a nut or acorn was that year to be seen in the neighborhood. They consequently flew so high, that different trials to reach them with a capital rifle proved ineffectual; nor did the reports disturb them in the least. I cannot describe to you the extreme beauty of their aërial evolutions, when a hawk chanced to press upon the rear of the flock. At once, like a torrent, and with a noise like thunder, they rushed into a compact mass, pressing upon each other towards the center. In these almost solid masses, they darted forward in undulating and angular lines, descended and swept close over the earth with inconceivable velocity, mounted perpendicularly so as to resemble a vast column, and, when high, were seen wheeling and twisting within their continued lines, which then resembled the coils of a gigantic serpent.

Before sunset I reached Louisville, distant from Hardensburgh fifty-five miles. The pigeons were still passing in undiminished numbers, and continued to do so for three days in succession. The people were all in arms. The banks of the Ohio were crowded with men and boys, incessantly shooting at the pilgrims, which there flew lower as they passed the river. Multitudes were thus destroyed. For a week or more, the population fed on no other flesh than that of pigeons, and talked of nothing but pigeons. The atmosphere, during this time, was strongly impregnated with the peculiar odor which emanates from the species.

It is extremely interesting to see flock after flock performing exactly the same evolutions which had been traced as it were in the air by a preceding flock. Thus, should a hawk have charged on a group at a certain spot, the angles, curves and undulations that have been described by the birds, in their efforts to escape from the dreaded talons of the plunderer, are undeviatingly followed by the next group that comes up. Should the bystander happen to witness one of these affrays, and, struck with the rapidity and elegance of the motions exhibited, feel desirous of seeing them repeated, his wishes will be gratified if he only remain in the place until the next group comes up.

It may not, perhaps, be out of place to attempt an estimate of the number of pigeons contained in one of those mighty flocks, and of the quantity of food daily consumed by its members. The inquiry will tend to show the astonishing beauty of the great Author of Nature in providing for the wants of His creatures. Let us take a column of one mile in breadth, which is far below the average size, and suppose it passing over us without interruption for three hours, at the rate mentioned above of one mile in a minute. This will give a parallelogram of one hundred and eighty by one, covering one hundred and eighty square miles. Allowing two pigeons to the square yard, we have one billion, one hundred and fifty millions, one hundred and thirty-six thousand pigeons in one flock.

Major William R. King, writing in 1866, told of a flock he saw one May on the Canadian side of the Niagara River in southern Ontario:

Hurrying out and ascending the grassy ramparts, I was perfectly amazed to behold the air filled and the sun obscured by millions of pigeons, not hovering about, but darting onwards in a straight line with arrowy flight, in a vast mass a mile or more in breadth, and stretching before and behind as far as the eye could reach.

Swiftly and steadily the column passed over with a rushing sound, and for hours continued in undiminished myriads advancing over the American forests in the eastern horizon, as the myriads that had passed were lost in the western sky.

It was late in the afternoon before any decrease in the mass was perceptible, but they became gradually less dense as the day drew to a close. At sunset the detached flocks bringing up the rear began to settle in the forest of Lake-road, and in such numbers as to break down branches from the trees.

The duration of this flight being about fourteen hours, viz, from four a.m. to six p.m., the column (allowing a probable velocity of sixty miles an hour, as assumed by Wilson), could not have been less than three hundred miles in length, with an average breadth, as before stated, of one mile.

During the following day and for several days afterwards, they still continued flying over in immense though greatly diminished numbers, broken up into flocks and keeping much lower, possibly being weaker or younger birds.

John Muir, describing his teenage years in Wisconsin in the 1850s, wrote that:

It was a great memorable day when the first flock of Passenger Pigeons came to our farm, calling to mind the story we had read about them when we were at school in Scotland. Of all God's feathered people that sailed the Wisconsin sky, no other bird seemed to us so wonderful. The beautiful wanderers flew like the wind in flocks of millions from climate to climate in accord with the weather, finding their food – acorns, beech-nuts, pine-nuts, cranberries, strawberries, huckleberries, juniper berries, hackberries, buckwheat, rice, wheat, oats, corn – in fields and forests thousands of miles apart. I have seen flocks streaming south in the fall so large that they were flowing over from horizon to horizon in an almost continuous stream all day long, at the rate of forty or fifty miles an hour, like a mighty river in the sky, widening, contracting, descending like falls and cataracts, and rising here and there in huge ragged masses like high-splashing spray. How wonderful the distances they flew in a day – in a year – in a lifetime!

In the autumn of 1847, in Kentucky yet again, but this time near Hartford, Benedict Revoil noted:

... on emerging from the wood, I observed that the horizon was darkling; and that after having attentively examined what could have caused so sudden a change in the atmosphere, I discovered that the clouds – as I had supposed them to be – were neither more nor less than numerous flocks of pigeons. These birds flew out of range … so I conceived the idea of counting how many troops flew over my head in the course of an hour. Accordingly, I seated myself tranquilly; and drawing from my pocket pencil and paper, I began to take my notes. In a short time the flocks succeeded each other with so much rapidity that the only way I could count them was by tracing manifold strokes. In the space of thirty-five minutes, two hundred and twenty bands of pigeons had passed before my eyes. Soon the flocks touched each other, and were arrayed in so compact a manner that they hid from my sight the sun. The ordure of these birds covered the ground, falling thick and fast like winter's snow …

An arithmetician of the district made a sufficiently curious approximative calculation of the number of individuals composing these extraordinary legions … Taking for example, a column about five hundred yards in breadth – which is much below the ordinary measurement – and allowing three hours for the birds composing it to accomplish their flight, as its swiftness was five hundred yards a minute, its length would be two hundred thousand yards. Supposing now, that each square yard was occupied by ten pigeons, we may conclude that their total number amounted to a billion, one hundred and twenty millions, one hundred and forty thousand.

This is an interesting account, but one which would not score that highly, as Schorger pointed out, for biological or mathematical accuracy. Five hundred yards per minute is a flight speed of only 17 miles per hour (27 km/h), about a third of what we believe to be the Passenger Pigeon's actual speed. But in the above account there is also some mathematical error or misunderstanding between Revoil and his 'arithmetician of the district' – for using the figures of a three-hour passage of birds in a column 500 yards wide and with the high density of 10 birds per square yard of ground covered (though flocks will have differed in structure, and there are plenty of reports of layer upon layer of birds in the air) then I calculate that there may have been 450 million birds in this flock if they were flying that slowly, but more than 1.5 billion if they were rattling along at 60 mph. The account is vaguer than many others, so it is difficult to be sure.

W. J. McGee wrote in 1910 of his recollections of birds in eastern Iowa in the 1860s and 1870s:

> A rough estimate of the number of birds passing a given point in spring may be useful. The cross-section of an average flock was, say, a hundred yards from front to rear, and fifty yards in height, and when the birds were so close as to cast a continuous shadow there must have been fully one pigeon per cubic yard of space, or 5,000 to each linear yard of east–west extension – i.e. 8,800,000 to the mile, or (with reasonable allowance for the occasional thinning of the flock) say 30,000,000 for a flock extending from one woodland to the other. Since such flocks passed repeatedly during the greater part of the day of chief flight at intervals of a few minutes, the aggregate number of birds must have approached 120,000,000 an hour for, say five hours, or six hundred million pigeons virtually visible from a single point in the culminating part of a single typical migration.

Schorger quotes the Allen brothers, who lived in New York State up until 1854, thus:

> It would be hard to make any estimate of their numbers that people would believe at this late day. I was going to say that a thousand million could have been seen in the air all at once. There would be days and days when the air was alive with them, hardly a break occurring in a flock for half a day at a time. Flocks stretched as far as a person could see, one tier above another. I should think it would be safe to say that millions could be seen at the same time.

These accounts of flocks are a selection from the many accounts that exist – chosen in some cases for their lyrical and descriptive nature and in others because they represent the most interesting examples of attempts to estimate the size of flocks or the numbers of birds passing a particular point over a period of days.

Reviewing the accounts as a whole, there are recurring themes such as the noise that the birds' wings make as huge flocks pass overhead and the darkening of the sun as the birds pass across the sky in such huge numbers.

The three most specific estimates of the size of a Passenger Pigeon flock are as follows: more than one billion by Audubon, more than two billion by Wilson, and more than three billion by Schorger interpreting King's observations. These three merit further scrutiny. All of them depend on estimating the area of the flock and the density of birds within it.

Crucial to the estimate of flock area is the speed at which Passenger Pigeons fly, as the estimate depends on multiplying the time that it took a flock to pass by the speed of the birds, to arrive at the length of the flock. These three key estimates of individual flock sizes all use a flight speed of 60 miles per hour (97 km/h), which is by no means the fastest estimate of Passenger Pigeon speed, given that different observers' figures range between 40 mph and 100 mph. Estimating the speed of a bird in flight is a tricky thing to do, and if this figure were badly wrong then so would be the estimates of flock size.

Almost all accounts of Passenger Pigeon flocks give the impression of swiftness – immense numbers moving at immense speed. No accounts suggest that a moving flock is dawdling or hanging around. Mourning Doves, Barbary Doves and Turtle Doves have all been measured flying at speeds of 50–60 mph (80–97 km/h), and the Passenger Pigeon was a slightly larger bird with very noticeably large flight muscles. Homing pigeons will maintain an average flight speed of 80 km/h over distances of 800 kilometres, and can average getting on for a mile a minute (95 km/h) over shorter but non-trivial distances (160 km).

In Michigan, field measurements were made as Passenger Pigeon flocks passed over 1-mile (1.6 km) sections of land, with men at either end signalling the arrival and departure of the flock in a section by waving white flags. Land was settled in many parts of the Midwest of the USA in a grid system of 'townships' consisting of 36 1-mile square 'sections' – so the landscape was arranged well for this type of measurement, and the populace were also well accustomed to judging distances of around a mile. These measurements gave flight speeds of around 60 mph. A speed of around 60 mph for the 'Blue Meteor' seems eminently reasonable.

Wilson's estimate of the numbers in the single flock he saw in Kentucky one summer in the early nineteenth century (more than two billion) depended on an estimated flight speed of 60 mph, a passage of birds of four hours, a flock width of a mile and a density of birds of three per square yard. He then does the maths correctly: 4 (hours) × 60 (mph) × 1760 (yards/mile) × 1760 (yards in width) × 3 (birds per square yard) = 2,230,272,000. Wilson described this estimate of numbers as an 'inconceivable multitude', but also said that it was probably 'far below the actual amount'. From what Wilson wrote, a duration of four hours and a flock width of a mile were both likely to be underestimates rather than overestimates, and so we could regard this flock as likely to have numbered between two and three billion birds. However perfect this estimate of this one flock might be, it is unlikely that Wilson watched all the Passenger Pigeons alive in the USA and Canada on that day in Kentucky. Was this flock half, or a quarter, or even a tenth of the total population of the species at that time? Even allowing for the fact that there might (or might not) have been some recently fledged young in the flock, and even dividing by two to arrive at pairs rather than individuals, the overall Passenger Pigeon breeding population of the USA and Canada must, surely, have been more than five billion pairs. To believe otherwise is either to question the veracity of Wilson's account, which is perfectly reasonable except that it seems to be backed up well by other estimates that we will soon examine, or to assume that he really was, on that summer's day in Kentucky, looking at nearly all the Passenger Pigeons in the world – despite the species having a breeding range which extended over hundreds of thousands of square kilometres.

Audubon made his estimate in a similar way to that of Wilson: 3 (hours) × 60 (mph) × 1760 (yards/mile) × 1760 (yards in width)

× 2 (birds per square yard) – and since he used a flight time of three hours and a density of two birds per square yard his estimate is exactly half that made by Wilson a few years earlier and a few kilometres further east in Kentucky. Audubon's was an autumn flock of an estimated 1,115,136,000 birds.

However, Audubon saw numbers of Passenger Pigeons on his journey between Henderson and Louisville for three days. It is clear that Audubon's estimate was for just one flock, as that is what he says at the beginning and end of the paragraph in which the estimate is given. He mentions that the birds were passing from northeast to southwest. He also says that the birds passed in 'undiminished numbers' for three days. The account sounds more like a description of a three-day passage of birds in a single direction than the same birds heading hither and thither over a three-day period, though we cannot be certain of this. But in any case, Audubon's estimate of a single flock flying for three hours does not appear to be an estimate of all the Passenger Pigeons that streamed over the heads, and mostly out of range of the guns, of the people of Kentucky and (on the other side of the Ohio River) Indiana over those three autumn days in 1813.

Audubon's estimate was of birds passing at more than 370,000,000 birds an hour, and he says that they passed for three days. He first mentions seeing them on his journey that day on the eastern side of Hardensburgh (now spelt Hardinsburg), and as he travels onwards he mentions the noonday sun being shaded by their passage. The birds are still passing overhead as Audubon stops for 'dinner' (was this perhaps a late lunch?), and presumably to rest his horse, at Young's Inn on the Ohio River. They flew in a 'front reaching far beyond the Ohio on the west, and the beechwood forests directly on the east of me', and when Audubon reached Louisville, before sunset, the birds were still passing by. Audubon gave the distance between Hardinsburg and Louisville as 55 miles, which is a slight underestimate, but throughout that journey it seems he was seeing Passenger Pigeons in enormous numbers all the time.

Let us just imagine that over the course of the three days pigeons passed at the rate that Audubon noted for a total of 12 hours – that would be around 4,500,000,000 birds. And it is not at all clear, considering that the birds were passing on a wide front, that the numbers could not have been greater, perhaps much greater. This was an autumn flock, so it would have contained birds of the year, but if there were 4.5 billion birds we would expect about three billion of them to have been adults, which would have amounted to 1.5 billion breeding pairs a few months earlier in the year. And again, were these all the Passenger Pigeons in the world, all seen by Audubon on just one three-day trip? It seems unlikely – but what factor should one use to multiply up?

Schorger uses King's observations to estimate the size of the flock that he observed on the Canadian side of the Niagara River as follows: 10 (hours) × 60 (mph) × 1760 (yards per mile) × 1760 (yards in width) × 2 (birds per square yard) = 3,717,120,000 birds. King states that the flock was a mile wide (funny how they so often were!) and that the birds were passing by from 4 am to 6 pm on the first day and in smaller numbers (but still 'immense' numbers) for another two days, so Schorger's use of 10 hours of flight is, as he recognised, on the conservative side. A doubling of this number, which would be perfectly reasonable given King's account, would bring the estimate to around 7.5 billion birds. This flock, or series of flocks, was flying from east to west into Canada in spring, presumably to nest. This may have been more than 3.5 billion pairs of Passenger pigeon heading into southern Canada to nest. We do not know the year of the observation, but it is likely to have been a little later than those of Wilson and Audubon, and it was described in King's book of 1866. This last of the 'big three' flock estimates was also, arguably, the largest. And again, it is unlikely that this flock comprised all the Passenger Pigeons in the USA and Canada at the time.

These three accounts are really the best ones from the point of view of gaining some rough and unreliable estimate of the flock sizes. I am left in no doubt that thousands of Americans and Canadians living through the sixteenth to the nineteenth centuries were blessed, and I think that is a fair description, by seeing billions of Passenger Pigeons passing overhead as they worked in the fields or went about their business in the towns. These were shared experiences of those living in Iowa and New York, and Tennessee and Quebec. The accounts of the time were not concerned with the population numbers of this species – they were simply describing an amazing natural phenomenon. They described rivers of arrow-shaped birds passing at haste through the skies in coordinated and busy flocks, coming from the far distance and heading into the opposite far distance on their migrations or feeding movements. And occasionally your neighbourhood might be chosen for a vast winter roost or breeding colony involving millions and millions of birds.

THE POPULATION SIZE CONUNDRUM

How do we, at this distance in time, make any sensible estimation of Passenger Pigeon numbers?

In winter the Passenger Pigeon seems to have ranged very widely, occurring from Hudson's Bay in the north (in December, which must certainly count as winter at that latitude) down to Texas and northern Florida, and from the Atlantic coast into the Great Plains. The winter roosts were vast smelly assemblages of birds but there are no useful estimates of the numbers of birds using individual roosts, and even if there were we have no information on how many roosts there might have been. We can imagine that many of these roosts held millions and probably tens of millions of birds, but that is only guesswork – and with no figure at all for the number of roosts, winter numbers are not amenable to estimation.

The accounts of Audubon, Wilson, Bendire and Cooper paint pictures of the extent of the colonies over tens and hundreds of square kilometres, and tell us that individual trees held scores, perhaps sometimes even hundreds, of nests. These colonies were on a scale that the world has not seen since, and yet, again, these impressive accounts don't enable us to reach any sensible estimate of total population numbers, since we don't know how many birds there were per area of colony and we don't know what area the Passenger Pigeon colonies of North America covered.

And so we are left with the flocks – and primarily with the accounts of Audubon, Wilson and King, because they give the most information – although their observations are in line with the less detailed accounts given by many others. All three accounts occur in the observers' published books, and none of them is given particular prominence. Reading each of these works, it seems to me that they give a believable account of North American wildlife; none is full of ridiculous stories and none seems a hotbed of exaggeration, and therefore it would surely be unkind and unreasonable to assume that the passages concerning the Passenger Pigeon were deliberately exaggerated for effect.

So might the three observers have been honestly mistaken in their estimates? Of course they might have been, but I can see no reason why they might not have honestly underestimated the numbers, just as easily as they might have honestly overestimated them. Perhaps 60 mph is towards the very top end of the expected speed of flocks, but it is pretty well supported by apparent field measurements at the time, and by studies of similar species much more recently. Perhaps it is a little odd that all three observers used a distance of a mile as the width of their flocks – did Audubon and King copy Wilson, perchance? But again, it is as likely that this was an underestimate as an overestimate.

It seems very likely to me that first Wilson (in summer, in the first decade of the nineteenth century, in Kentucky), then Audubon (in 1813, in autumn, in Kentucky) and then King (in May, sometime between 1830 and 1860, in southern Ontario) all saw flocks of Passenger Pigeons that numbered in the billions. And so, quite possibly, did many others who gave less detailed accounts of their sightings.

Unless all the Passenger Pigeons of North America had gathered into single flocks for the benefits of these observers, which seems very unlikely indeed, these sightings represent an unknown proportion of the total American Passenger Pigeon population. And they were made less than a century before the Passenger

Pigeon was driven to extinction, but in a period when it had probably already declined fairly markedly (see Chapter 5).

Schorger 'plumped' for a population of around three billion Passenger Pigeons at the time of the beginning of the European invasion of North America. I'm not sure that he felt that comfortable about it as he wrote, 'As a guess, I would place this population at 3,000,000,000 at the time of the discovery of America, with a possibility of 5,000,000,000.' My guess would be higher: I would plump for an initial population, before European interference, of between five and ten billion birds.

There is one check that we can make on this guess – it's not a very strong check, but worth making all the same. If there were, let's say, five billion Passenger Pigeons abroad in the forests of northeastern North America in the sixteenth century, would there have been enough food to feed them?

Here is an admittedly rough-and-ready calculation. A free-living Passenger Pigeon would need about 50 grams of food per day to survive (based on Mourning Dove physiology and known food intake of captive Passenger Pigeons). So five billion Passenger Pigeons, for a year, would require about 9×10^{10} kilograms.

The feeding of the five billion would, during the course of the whole year, be spread over a total distribution of about 3.5×10^8 hectares. So how likely is it that North American forests, in pristine condition, might produce an average of 260 kilograms of tree mast each hectare each year?

Quite possible, it seems. American Beech can produce up to 2,800 kilograms of mast per hectare – but they only do it two or three times every eight years or so. Oaks produce lower peak mast amounts, around 700 kilograms per hectare, but they do this more frequently than beech. And American Chestnut is a more regular performer at around 400 kilograms per hectare.

An attempt to estimate the mast production for forests in North Carolina that were dominated by oaks and chestnuts suggested that average total mast production would be more than 400 kilograms per hectare per year, and in the best years upwards of 700 kilograms per hectare.

The data on mast production aren't great – for one thing they weren't measured from the forests of eighteenth-century North America in the absence of human interference, and for another they aren't taken from right across the range of the Passenger Pigeon, but as a rough estimate it appears that there would have been enough mast produced to feed billions of Passenger Pigeons, although these (very rough) calculations suggest that the pigeons might not have left much food over for other species.

Passenger Pigeons may have eaten a very high proportion of the mast production of their favoured tree species each year – that's hardly surprising, as there were billions of them! A population of around five billion is not completely unreasonable, according to these rough estimates, and even 10 billion would not be out of the question if there were a run of years of high productivity for trees that led to high productivity and survival for Passenger Pigeons. Perhaps, in some autumns, the Passenger Pigeon population topped 10 billion. I'd like to think so, anyway.

THE PASSENGER PIGEON RESEARCH PROGRAMME, PART 2: TO RESOLVE THE CONUNDRUM OF NUMBERS

If the Passenger Pigeon were alive today it would be a difficult species to survey accurately – a highly colonial species, but which nests in different places in different years, is a challenge for any survey.

In both North America and the UK, well-designed bird surveys, capable of measuring year-to-year population trends, started in the early to mid-1960s. In both the USA/Canada and UK/Ireland these surveys have evolved over time, but both are now called the 'Breeding Bird Survey'.

The US and Canadian scheme uses standardised car routes of 24.5 miles (39 km) with frequent stops to count birds, whereas the UK/Ireland scheme requires walking two parallel transects of 1 kilometre each. The North American scheme has more than 4,000 routes, and the UK/Irish scheme has more than 3,000 randomly chosen 1-kilometre squares. These sites are covered in the spring in order to monitor the trends in numbers from year to year.

This approach is perfectly adequate for many species of bird: the warblers, thrushes, finches and sparrows that make up a large part of the birdlife of an average walk in the countryside (or drive if you are American – sorry, couldn't resist it). However, when I submit my BBS data each year I am asked whether there were any colonial species and whether I counted their nests. There aren't in the areas that I cover, and the only likely species would be Rook and Grey Heron.

Maybe a quarter of American BBS routes would be within the breeding range of the Passenger Pigeon, but only a tiny proportion of them would coincide with nesting colonies. Imagine the prospect of a modern-day BBS counter faced with a flock like one of those encountered by Audubon and Wilson and having to count it. No, a different approach would be needed, although the BBS would be very useful for getting a much better estimate of the number of Passenger Pigeons nesting in a dispersed manner away from colonies and over a much wider range – that is the type of bird for which the BBS is well-adapted.

But for Passenger Pigeons nesting in huge colonies, the task is much more like that of estimating numbers of colonial seabirds – estimate the number of colonies and their size. These days, with modern communications, locating the colonies, at least the very largest of them, would be easy enough – you wouldn't easily miss a few million pigeons if they lived just down the road from you.

Given the short life of any single colony, around a month, it would be important to synchronise survey effort. The best method would probably be to use coordinated aerial surveys to locate the colonies and estimate the extent of each. Ground-based visits to colonies would then be used to measure the density of nests in several sample areas, and the population size could be calculated by multiplying total area of colonies by average density of nests.

Estimating the population size of a modern-day Passenger Pigeon population would still be a tough nut to crack (rather appropriately, for a bird feeding on tree mast), but crack it we could. Unfortunately (unfortunately for it, and also for our inquisitiveness), the Passenger Pigeon ceased to be, long before any such methods could be used.

BILLIONS OF THOUGHTS OF ABUNDANCE

Whether the population was three billion, five billion or ten billion, or more, it is difficult to picture such abundance, and I think that is one reason why we tend to aim low in our estimates; we cannot imagine that the forests of pre-invasion America were so rich in wildlife. We've never seen it and we can't picture it in our minds – our imaginations can't cope. And I suspect that we feel guilt too – could we really have caused the extinction of a bird that was so numerous?

Conservationists talk of the 'shifting baseline syndrome', a term coined by the American marine biologist Dan Pauly. Pauly was considering the situation where fisheries biologists wish to restore a depleted stock, and his argument was that we tend to restore things back to what was familiar to us in the past rather than to a much earlier state of richness. Each generation tends to work within the cramped confines of its own remembered baselines.

In the case of the Passenger Pigeon we find its previous huge numbers difficult to imagine. Because it has been gone for so long we have seen nothing like it. We do not see a small number of Passenger Pigeon colonies and wish there were more, or see colonies of hundreds of thousands of birds and wish there were still millions – we have lost the collective memory of the commonest bird on the planet. Gone are the times now, as Aldo Leopold wrote, when there were old men who could remember the Passenger Pigeon in their youths. With the centenary of Martha's death those times are long gone.

When we lost the Passenger Pigeon from Earth, then, we lost an abundant and unique species. We cannot look at the Mourning Dove or the Wood Pigeon and see a species like the Passenger Pigeon in either abundance or ecology. Yes, there are other pigeons, but there is none that truly resembles this forest pigeon that once darkened the skies of North America.

We talk more often these days of billions in terms of financial crisis, and usually in terms of debt, but for a century now we have been in ecological debt to the tune of billions of Passenger Pigeons. But a billion (1,000,000,000), one thousand million, is a number beyond the grasp of our minds. Here are a handful of thoughts that may help you to approach an appreciation of how abundant the Passenger Pigeon was.

The current human population of the world, at the moment I am writing these words, is estimated to be more than seven billion; in fact it is 7,214,589,475, but it keeps going up! When Alexander Wilson reached the age of 38, in 1804, and four years before he began to publish his great ornithological work, the human population of the world reached one billion, of which only about six million were found in the USA. When he calculated that the flock of Passenger Pigeons he watched in central Kentucky was around two billion birds, that was more than 300 times the human population of the USA and twice the world population on that day.

The First World War began just over a month before Martha died. No-one quite knew what a bloody affair it would be; at its end it ranked fourth bloodiest in human history in terms of military and civilian losses (now relegated to fifth after World War II). World War I saw the deaths of 16 million military souls in its four years – four million combatants each year. In comparison, the war against the Passenger Pigeon must have despatched at least two billion birds in a century, which is a death rate of 20 million pigeons a year for a century – a death rate five times as high over a period 25 times as long.

The first day of the Battle of the Somme, 1 July 1916, claimed the lives of nearly 20,000 British soldiers. The last century of the long decline of the Passenger Pigeon took out two Somme-sized death tolls from its population every day – every day for a hundred years. The scale of loss and the rate of loss are both staggering. History shows that humans can be very unkind to other humans, but also very, very unkind to nature.

Schorger suggested, on the basis of his estimate of 3–5 billion Passenger Pigeons, that around a third of the biomass of all c.800

species of North American birds might at one time have been
Passenger Pigeons (despite their rather restricted range). Maybe it
is only fitting that the Peterson *Field Guide* retains that image of a
Passenger Pigeon looking accusingly out at us from page 181.

Today, in the UK, an area the size of a small American state,
there are approximately 84 million pairs of birds (including all
wild species). In Europe as a whole, nowadays, there are between
one and two billion pairs of birds – perhaps three billion individual
birds – in an area of around 10 million square kilometres. The
Passenger Pigeon was so abundant that its numbers were around
those of the total bird population of the European continent but
fitting into an area of North America, in the breeding season, only
one fifth that size. It's as though we turned all the birds of all the
species in Europe into Passenger Pigeons, and then pushed them
all into western Europe.

Some time on 1 September 2014, between midday and 1 pm,
will mark the exact centenary of the death of the last Passenger
Pigeon on the planet. Any clocks which have faithfully marked the
seconds since Martha's death have counted, a second at a time,
more than three billion seconds to arrive at the centenary of
Martha's death. A million seconds only reached to 13 September
1914 – less than a fortnight. A billion seconds were passed in
March 1945, and 3,155,760,000 (give or take a few) will be
reached on the exact centenary. A century of counting, second by
second, backwards through the days and through the nights from
the time of Martha's death, would take us back to the time when
Audubon and Wilson were seeing their flocks of Passenger
Pigeons. And ten billion seconds would reach the 1690s – the time
of the Salem witch trials, a decade or so after Pennsylvania was
founded, and when the human population of the European invaders
on what would become US soil was less than a quarter of a million.

I've seen a lot of flocks of birds in my time. I have 'ooh-ed' and
'ah-ed' on the shores of the Wash as a quarter of a million Knot (I
was told) have wheeled in the sky, and I have gasped in the
dwindling light of dusk as somewhere around a million (I was told)
Starlings performed aerial manoeuvres in front of the setting sun

before settling to roost in a reedbed on the Somerset Levels. But I have never seen, and neither have you nor any other living person on Earth, a sight that gets close to the spectacle of a flock of Passenger Pigeons or their huge colonies or winter roosts.

Their abundance was phenomenal – they were one of the natural wonders of our world, and they are gone. We are left with what Aldo Leopold called 'book pigeons', which cannot dive out of a cloud or clap their wings in the mast-laden woods. I felt, as a producer of 'book pigeons', that I had to see the places which were important to the Passenger Pigeon, stand in a fragment of ancient forest which would have known the sky darken when the pigeons passed by, and visit the spot where Martha died. I wanted to talk to modern-day Americans about Passenger Pigeons and try to picture these birds back in their former haunts, and to see the places where they were slaughtered in their millions. And so, in May 2013, I set off for the former haunts of *Ectopistes migratorius*.

CHAPTER FOUR

A road trip in search of an extinct species

Here is your country. Cherish these natural wonders, cherish the natural resources, cherish the history and romance as a sacred heritage, for your children and your children's children. Do not let selfish men or greedy interests skin your country of its beauty, its riches or its romance.

Theodore Roosevelt

17 May 2013

Crossing the Mississippi River from Missouri into Kentucky marked the real start of my road trip, even though I had already driven nearly 700 miles from Texas. American literary and popular culture is peppered with road trips of different sorts, from Jack Kerouac's *On the Road* to Hunter S. Thompson's *Fear and Loathing in Las Vegas*, and from Callie Khouri's *Thelma and Louise* to John Steinbeck's *Travels with Charley*. There are even several natural-history road trips – from Roger Tory Peterson's and James Fisher's travels through *Wild America*, repeated by Scott Weidensaul in *Wild America Revisited,* to the less well-known journeys of Frank Fraser Darling in *Pelican in the Wilderness* and John Muir's *A Thousand Mile Walk to The Gulf*. But none was a road trip in search of a long-extinct bird.

You might question the value of travelling around the USA in search of inspiration about Passenger Pigeons, but the places that were important in the story are still there and I wanted to see them and get a feel for them. I wanted to see where the last Passenger Pigeon was shot in Pike County, Ohio, and to see that same mounted bird in the museum in Columbus, Ohio. I wanted to drive through the area of the biggest Passenger Pigeon nesting colony ever recorded and get a feel for its immensity.

But more than that, I wanted to see today's USA and imagine whether the Passenger Pigeon, if it were not extinct, could fit in to the present-day world. I wondered whether I would meet people who knew about Passenger Pigeons, and missed them, or whether there would be just blank faces whenever the bird was mentioned. And so I would talk to people, see places, learn about the forests and imagine the past to understand the present. So this is a road trip with fewer drugs, less sex, fewer shootings and fewer car chases than any of the others – and with a different purpose.

The Mississippi is a mighty river. When I was at school we learned two things about the Mississippi: that it was the longest river, or one of the three longest rivers, in the world, and how to spell Mississippi. Quite why either was supposed to be of great value to a Bristolian I don't know. The Mississippi has now dropped down to Number 4 of the world's rivers but I can still remember how to spell its name (Missus M, Missus I, Missus S, S, I, Missus S, S, I, Missus P, P, I).

I crossed where the Mississippi and Ohio rivers merge, between Sikeston, Missouri, and Wickliffe, Kentucky – the Ohio bringing water from states as distant as New York and Pennsylvania to mix with waters brought by the Missouri ('too muddy to drink but too wet to plough') from Canada and Montana as well as the Mississippi's own load from Minnesota and Wisconsin.

The Mississippi River traditionally forms an ecological boundary. To the east of it, where I was going, the natural vegetation of the land was primarily woodland, whereas to the west, behind me, were the Great Plains. And entering Kentucky as I did, points south of here were rare breeding grounds for the Passenger Pigeon, but were the lands it sought in winter, while the area to the north was where the bird nested.

I was heading for Henderson, Kentucky – a former home to a French draft-dodger whom the Americans took to their hearts.

There is nothing that appealing about Henderson at first glance. It resembles many small towns in the USA, with a string of fast-food restaurants advertising their wares in neon signs in the fading light of dusk. As you drive into town past the McDonald's, the

Burger King and the Taco Bell you could be almost anywhere in the USA – and, as is true almost anywhere in the USA, there is a motel looking for your custom.

The Henderson Downtown Motel was the only obvious motel in town, and it was pretty cheap and claimed to have been renovated recently. I booked in for two nights and stressed that I needed the wifi to work in my room and was assured that it would. When I tried it, it didn't, which meant an immediate return to the motel office, a conversation with Mr Patel the manager, and a transfer to a room nearer to the office where the wifi did work. By now it was getting dark and Chimney Swifts and Common Nighthawks were circling above my head. The jet lag was beginning to bite so it was time to get some sleep.

18 May 2013

John James Audubon is arguably the most famous name in American ornithology. His name was taken by the USA's foremost bird conservation organisation, the Audubon Society, and he has towns, roads, parks, a warbler, a shearwater, an oriole and, in Henderson itself, I noticed, a pawn business, named after him. Audubon Loans and Pawn is interested in your jewellery (or actually your jewelry – for we are now in America) and your guns.

Audubon was born in what is now Haiti in 1785 and moved to France as a youth but came to the USA in 1803 (to avoid the Napoleonic wars), becoming an American citizen in 1812. He first lived in Mill Grove near what is now Audubon, Pennsylvania, and then moved west to Kentucky. He had a love of nature and a skill in drawing. At Mill Grove he caught Eastern Phoebes and tied coloured threads to their legs – apparently the first bird-banding carried out on the American continent – and showed that the same birds return to their nest sites in successive years.

Audubon experienced the ups and downs of business, and the emphasis of his career moved gradually towards drawing and painting wildlife, most especially birds. He met the older Alexander Wilson (of warbler, phalarope and petrel fame) in Louisville, Kentucky, in 1810. When Wilson showed Audubon two volumes of his *Ornithology* Audubon would have subscribed had not his business partner, in French, told him that his own efforts were far better. Audubon apparently showed his own work to Wilson, who was relieved that Audubon seemed to have no plans to publish it. But this meeting may have encouraged Audubon to work towards the publication of what eventually became *The Birds of America* with its 435 life-sized plates of America's birds. If so, he suffered a serious setback in 1812 when his collection of drawings was eaten by rats – but at this time business was good and continued to be so until a little before 1819, when Audubon went bankrupt. Audubon wrote of the time when he was based in Henderson with great affection, 'The pleasure I felt at Henderson and under the roof of that log cabin can never be effaced from my heart until after death'.

East of here, on his way to Louisville from Henderson, at a place called Hardinsburg, Audubon saw the flock of Passenger Pigeons whose number he estimated at more than a billion. Hardinsburg is still there even though there aren't a billion pigeons any more. I set off from Henderson early in the morning in the drizzle. Driving through the agricultural landscape with scattered woodland, it didn't look very interesting.

Hardinsburg was a small place, only about 30 years old, when Audubon was there. Today, John Deere and New Holland compete to sell you a tractor, across the road from each other, as you enter town. There is also a US Department of Agriculture Service Center, a Farm Bureau and a Farm Credit Services. Hardinsburg is a farming town.

I looked for breakfast in drizzly Hardinsburg. Breakfast is the most important meal of the day for a traveller in the USA. Not only is it arguably the most American in content but it is also the most social in nature, and therefore you see the real America more through breakfast than through lunch or dinner. Americans are much more likely to eat breakfast out than are the British, and so the local diner, café or restaurant becomes the place where news is exchanged.

There was nowhere obvious for breakfast in my first drive

around the town so I got some gas from the Marathon station at the crossroads and asked Trish where I should look. After trying to sell me a microwaved sausage in a roll she told me where to find Jake's Place. Trish had heard of Audubon but not of Passenger Pigeons. She was very helpful but not a waitress.

Susan is a waitress. She is short, thin, a bundle of energy, quite talkative (with the confidence that only comes from having spent a few decades on this planet) and limps in her right leg. I ordered the special (since this was the first food since breakfast yesterday) of two eggs over easy, hash browns, ham, toast and coffee. Susan persuaded me to have white gravy with it, which is made of every type of fat you can imagine churned up together.

At the other end of the room a bunch of men, mostly in baseball caps, were chatting. They were clearly locals and felt comfortable about helping themselves to more coffee from the jug on the coffee-maker. I was sitting at the quiet end of the room but got chatting to two guys from Pennsylvania about Passenger Pigeons. They were amazed that anyone should come to this part of Kentucky because of a long-extinct bird (of which they had never heard) but they wished me luck in my travels. Susan told us that she had had a Passenger Pigeon in her garden once. Apparently she'd also had 'that extinct woodpecker' in her garden. I wish I'd had an invite!

I liked Susan. She was full of energy, she was working hard, and yet she still had conversations with lots of people, including me. My suspicion is that Susan is a bit interested in birds but wouldn't admit it. She asked whether I was here to try to prove that the Passenger Pigeon wasn't extinct but I told her 'no' and that it was surely long gone, but this was where an enormous flock had once been seen, a couple of centuries ago. She asked whether I had enjoyed my breakfast and I said I had, very much, as I had, very much, but that I wasn't sure about the white gravy.

'No,' Susan said, 'I didn't think you'd like that.'

Audubon had left Hardinsburg and was heading for Louisville when he encountered his flock of Passenger Pigeons at The Barrens. No-one could tell me where The Barrens might be, so I doodled around the lanes, turning up dead-ends where dogs, leashed and unleashed, greeted my car with barking. I was seeing birds all the time as I searched for the site of the sighting of a now-extinct bird.

American Robins hopped Blackbird-like – for they are of course thrushes – on cut lawns. Eastern Bluebirds caught flies from fence-wires and returned to nest-boxes with their prey. Bright red Northern Cardinals surprised me with their colour, and Brown Thrashers flew across the road and dived into bushes. Just to make me feel at home, Starlings occasionally flew over carrying food. The sun now shone.

I came to a hill with thin soils and rocky outcrops which might well have been the type of place to be called The Barrens in the past. Now it had a few scattered houses and, on breasting the crest of the hill, there was a fine view over treetops to the northeast. Perhaps this was the spot where Audubon saw the Passenger Pigeons fly from northeast to southwest, and sat for a while to count their numbers. He found the counting impracticable as the birds passed by in multitudes and so he travelled on, with the air 'literally filled with pigeons' and the 'light of noonday … obscured as by an eclipse' while 'the dung fell in spots, not unlike melting flakes of snow; and the continued buzz of wings had a tendency to lull my senses to repose.'

As Audubon travelled on, so did I, roughly following his route until I reached West Point, Kentucky (not the site of the military academy, which is in New York). To get there, you pass by Fort Knox. You don't see much of the large army base, and you don't get so much as a glint of gold, yet alone a glimpse of any alien spaceships stored there. But, 200 years ago, Audubon stopped at West Point and watched the riches of the Passenger Pigeon fill the sky. 'Whilst waiting for dinner at Young's Inn,' he wrote, 'at the confluence of Salt River with the Ohio, I saw, at my leisure, immense legions still going by, with a front reaching far beyond the Ohio on the west, and the beechwood forests directly on the east of me.'West Point was less than 20 years old when Audubon visited, as it was established in 1796 after a peace treaty ended attacks by Native Americans. West Point got its name from being the westernmost point of 'English civilisation' on the Ohio when it was established by James Young, and his inn was a famous and profitable location for a good many years. As well as Audubon, American politicians from Andrew Jackson to Wendell Wilkie visited this inn, and the 'Swedish nightingale', Jenny Lind, once sang from its steps.

It took me a while to track down the inn. I had wondered whether there was still an inn and I could have lunch where Audubon had eaten and watched the stream of Passenger Pigeons, but a drive around failed to locate it and a couple of people I asked were unaware of its existence. A youngish man attaching his boat to a trailer hadn't heard of Young's Inn and I almost gave up, but thought I'd ask the elderly lady walking her dog a few yards further on – and she pointed 20 yards further on again and said 'Right there!'

Young's Inn is now a private house, a very nice one by the look of it, overlooking the Ohio River – and the view of the river from here can't have changed very much since Audubon's day. There is a historical marker on the lawn saying that it was built in 1797 by James Young as a stagecoach stop on the Louisville and Nashville turnpike, and that Audubon stopped here and wrote about Passenger Pigeons. A groundhog walked along the side of the house.

West Point is a quiet pretty place, and I thought I'd look around a little further afield. Crossing the Salt River and heading along the Ohio River north and east, you are almost immediately in a much more industrial landscape. As you pass up the Dixie Highway there is a faded billboard for the now-closed Doe Run Inn which used to offer a 'natural, historic, and romantic' stay. Less romantic, I suggest, is the strip joint of the Riverside Lounge which offers 'girls, girls, girls' and 'Lime-a-rita' and 'Straw-ber-rita' to drink; I didn't stop. A little further on are the three chimneys of the Mill Creek coal-fired power station of Louisville Gas and Electric on one side of the road and a CEMEX plant on the other. Clouds of steam have replaced clouds of Passenger Pigeons in these skies.

I needed to move on to see where Audubon's rival Alexander Wilson had seen his flock of Passenger Pigeons, on the other side of Louisville and between Shelbyville and the state capital of Frankfort.

Wilson was a Scot, from Paisley. He was born in 1766 and was thus Audubon's senior by 19 years. He is known, and with good reason, as the father of American ornithology, and although he was certainly a less accomplished artist than Audubon he was also certainly a more scientific observer.

Audubon and Wilson were probably too similar to get on well, and their friends did nothing to dampen down any rivalry between them. There were charges and counter-charges of plagiarism. Interestingly, it is only in the pages of their two works that you will find the Small-headed Flycatcher illustrated – no such species exists, nor is it obvious what real species either might have meant, and yet it appears in both their works and nowhere else. This appears to be an instance of Wilson having copied a painting that Audubon showed him, and yet there are other instances of where Audubon's composition (for instance, his Bald Eagle and Mississippi Kite) so closely resembles Wilson's earlier work that it is difficult to reject the idea that they were 'copied'. But we needn't bother ourselves with such ancient disputes, except to wonder what might have occurred if these two greats of American bird history had got together and pooled their resources.

Wilson's great work, *American Ornithology*, runs to nine volumes and was published between 1808 and 1814 – the last volume shortly after Wilson's death. It was the first to attempt to illustrate all of the birds of America, covering 268 species. Wilson's work was the pinnacle of American bird art until Audubon's first instalment of *The Birds of America* was published in 1827.

The area where Wilson saw his flock of Passenger Pigeons is a bit uninspiring – or at least it didn't inspire me. I travelled along the small roads and saw Grey Catbird and Spotted Sandpiper. Central Kentucky was enjoying itself outdoors in the sunshine and no Passenger Pigeons darkened the sky. I turned the car around and headed back to Henderson.

That evening I ate in the On Deck Riverside Bar and Grill – sitting outside in the warm air and watching the sun set over the trees on the far bank of the Ohio River. I ate tacos and drank not Coca Cola, nor Pepsi Cola, but Royal Crown Cola, another type of cola drink which is pretty indistinguishable from the other two when it is heavily iced and gulped down as much as a coolant as a thirst quencher. If only these three colas could distribute themselves geographically like America's three Bluebirds (Eastern, Mountain and Western) so that you knew which one to expect from your location.

Just down the road there is the old part of town, where it looked to me as though there might be a handful of buildings that Audubon would recognise if he were to return here. He would also recognise the Ohio River as it made its stately way west to join the Mississippi, but there is precious little else that would be recognisable after 200 years. When he lived here it was close to the frontier beyond which so-called civilisation had not penetrated.

But Audubon would recognise the Chimney Swifts and Common Nighthawks circling above me. In nearby Louisville he had visited a sycamore tree 7–8 feet (2.1–2.4 m) in diameter (which would make it quite a rarity these days) where Chimney Swifts roosted overnight. He watched them go in one evening before a thunderstorm, and returned early in the morning to see how they had fared. He estimated that about 9,000 swifts roosted in that tree that night.

Audubon noted that the nighthawk arrived in the Middle States (which I take to include Kentucky) around 1 May and then laid its eggs around 20 May – so I guess they were just about to lay as I watched them in the warmth of this May evening. He also noted that these birds alighted on trees or sometimes chimney-tops, from which they emitted their harsh notes – and that's just what they were doing at the Downtown Motel in Henderson tonight.

19 May 2013
It was Sunday, and although I didn't go to church I paid my respects to Nature (and Man, a bit) in four different ways.

I started with some early-morning birding in the John James Audubon State Park. The jet lag got me up early but I was very happy to have a few hours to stroll around the park and see some birds. On the lake there were Hooded Mergansers; on its shore, a Belted Kingfisher. The open area by the car park had Chipping Sparrow and American Goldfinch, and in the wooded plots I found Carolina Wren, Cedar Waxwing, Northern Parula, Tufted Titmouse, Carolina Chickadee and Eastern Wood Pewee.

When it looked like it might rain I nipped out to fill up with gas and buy a coffee. At the gas station I was told 'I love your accent' by the young cashier, and I replied that I loved hers too.

Back to the John James Audubon State Park for a spot more birding and a wait for the museum to open. I watched, and listened to, Carolina Wrens at close quarters, and I think I have their song fixed now. Although, to be fair, I had thought it was one that I already knew. 'Tea kettle, tea kettle, tea kettle' describes it quite well, but the first I heard today made me think 'thrush', not 'wren'. How many extra species would I record if I had 'American ears'?

Birding was my first way to pay my respects. Visiting the museum in the Park Nature Center was my second. The Nature Center has a shop (I bought postcards), a discovery room (where a father and an older brother were encouraging a younger brother's interest in nature – nice to see), a gallery downstairs which had an exhibition of local artists (I skipped that) and a viewing gallery over some bird feeders (stunning American Goldfinch) as well as the museum.

The museum is superb. At least it is if you are a fan of John James Audubon (and I am). You get his life history, an account of his rather frosty meeting with Alexander Wilson (I would have thought that a Frenchman and a Scot could at least have found common ground moaning about the English!) and a look at his art.

Audubon travelled widely collecting specimens and drawing birds. He made a living from painting portraits as he travelled, and his wife Lucy brought in the money while he was away through teaching. His attempts to get his work published in the USA were unsuccessful, so in 1826 he travelled to England, arriving in

Liverpool with letters of introduction – and at last his work received an enthusiastic reception. This woodsman from America, with his drawings of foreign birds, was a hit in England, and Audubon raised the money by subscription to get his work printed. The cost of printing the work as a double-elephant folio was around $2,000,000 in present-day money, and the money-raising was an immense achievement for a man most at home with his gun and pencil in the distant woods of the frontiers of the New World. But the achievement of pinning down, or stringing up, practically all of the USA's bird species, and illustrating them, was even greater.

The Birds of America was published as a loose-leaf series between 1827 and 1838. On 6 December 2010 a bound copy was sold at a Sotheby's auction for $11.5 million, a record price for a printed book. The museum at the John James Audubon State Park has four original double-elephant folio copies, each vast volume measuring 100 × 72 centimetres and depicting the birds at life size – and I saw them this morning.

Audubon's art gave European subscribers a wide window into the wildlife of the American continent and an insight into the avifauna of the New World. Some of the portraits are wooden and dull but others are simply breathtaking in their vibrancy, colour and beauty.

Audubon drew from skins but also from knowledge. His paintings are accurate, but the passerines, in particular, often look rather dead! However, many plates are incredibly beautiful: for example, the Wild Turkey, the Bald Eagle, the Great Blue Heron and, yes, the Passenger Pigeon. This was artwork celebrating nature, and I was pleased to go and celebrate Audubon in my own little way.

I had a very nice chat to the woman on the desk. We agreed that Audubon looked rather handsome, and I said he reminded me of the French footballer David Ginola (in the days when he was endorsing L'Oreal hair products), which got us onto David Beckham (who had just bowed to the inevitable and announced his retirement from the beautiful game), so it was time to leave.

Next stop was Cincinnati Zoo. As I approached Cincinnati on

the I71 the road made the left sweep down a steep hill and I saw the city, across the Ohio River in the state of Ohio, in all its glory. There were the skyscrapers in a huddle on the north bank and they were shining in the sun like a symbol of American dominance of ... well ... of pretty much everything really.

Finding the zoo on a previous visit in 2011 had been a nightmare – with satnav this time it was a doddle. I definitely went the quick way, and it definitely wasn't the same way as last time. The female voice of the satnav took me past the most priapic of the tall buildings and down Martin Luther King Avenue to the zoo.

The only change in two years was that it was now $15 to get in instead of $10, but still $8 to get out of the car park. It was a busy sunny Sunday afternoon and the zoo was packed with families, and a high proportion of them were African-American. I queued to get in behind a long line of visitors who were clearly going to make the most of the afternoon and see as many animals as they could. The last thing on their minds was an extinct pigeon, but that was the only reason I had come here.

I remembered the route to the Passenger Pigeon Memorial. It's easy to find – past Gorilla World and it's on your left.

A small pagoda marks the spot where Martha was caged for many years and where, with her last breath, her species came to an end. I shifted my feet, wondering where exactly she had died, but wherever it was I was certainly standing within a few feet of it.

There are mounted Passenger Pigeons and a net that was used to catch Passenger Pigeons, as well as a collection of twigs to represent a nest, and in it sit two white eggs (yes, two eggs – Schorger would not have been impressed). The small building has information about the fate of the Passenger Pigeon and the fate of its last representative on Earth – that female called Martha.

Martha is usually said to have been 29 years old when she died, a good age even for a captive and pampered pigeon. She was born in captivity and probably came to Cincinnati Zoo in 1902. The zoo itself had faced extinction a few years before her arrival. In January 1898 receivers were appointed to administer the zoo, and a year later they offered it for sale at an asking price of $90,000 but received no bids. In April 1899 it was announced that the Zoo had been bought for $75,000 – without which purchase I assume there would have been no home for Martha, here where I sat, and no reason for me to sit there a century later.

From April 1909 Martha had but one companion, named George (George Washington's wife and America's first First Lady was Martha). George died in July 1910, and for just over four years Martha was the only living member of her species.

She seems to have cut a forlorn figure, hardly moving in her cage for hours at a time. It was clear that the end of an individual bird, and with it the end of an abundant species, was fast approaching. The *Cincinnati Enquirer* reported that 'The days of the last Passenger Pigeon ... are now numbered' in its edition of 18 August 1914 – and two weeks later, on Tuesday 1 September, Martha died.

There is a bronze statue of a Passenger Pigeon by the memorial which I find very beautiful. It is smooth and a pleasing likeness of a pigeon. Its stone base carries a plaque with the following words: 'It is the hope of the Langdon Club that people who visit this memorial will want to work toward the preservation of all the world's fauna'. I do.

Before I left Cincinnati Zoo I spent some time sitting across from the memorial and watching the people go past – and go past they did. I counted more than 200 people walking past the memorial, and not one of them gave it so much as a second glance. Fair enough – let's concentrate on the living rather than the extinct.

As I left I went into the zoo shop to buy a baseball cap and some post cards. I would have bought a little memento of Passenger Pigeons if there had been any, but two shop assistants confirmed that in the well-stocked shop there was not a single item with a Passenger Pigeon on it. Retail managers are a canny bunch – they don't use shelf space for non-sellers, so that is further confirmation that it is only a strange Englishman who has travelled 3,000 miles and has been looking forward to visiting this zoo for months who would have parted with cash for a memory of Martha.

I imagine I was the only person who visited the zoo today,

perhaps this year, perhaps ever, who came to remember an extinct species and didn't look at a live animal.

So that was my third homage to nature.

I left Cincinnati and drove the 94 miles to Piketon, Ohio, on a sunny Sunday evening through Ohio countryside with lines of small hills running from north to south separated by wide flat valleys. The hills are mostly wooded with oaks and beech and the valleys are cultivated with corn and wheat. Piketon is small but my room in the Town and Country Motel was clean and cheap.

Just down the road was another location I was determined to visit, so as soon as I'd checked in I set out to find it. Chris Cokinos describes tracking down this location in his book, which I recommend, about extinct American birds: *Hope is the Thing with Feathers* (it's a line from an Emily Dickinson poem). Cokinos poked around in dusty rooms consulting dusty ledgers to discover the area of land where a teenage boy, Press Clay Southworth, shot the last wild Passenger Pigeon in March 1900. He'd done the hard work, and it was easy for me to follow the directions in his book to get to the area.

Driving south from Piketon on the 23, after just about four miles I turned left and crossed the railway line by 'Carter Kitchen and Bath' and then turned right to go down Wakefield Mound Road. I was keen to see where the Passenger Pigeon ceased, as far as we know, to be a wild bird. As I headed towards Wakefield it seemed to my English eyes a typically American scene of trailers and wooden houses surrounded by freshly cut lawns with the occasional Stars and Stripes hanging proudly in the yard. Admittedly there was a uranium enrichment plant over the brow of the hill – but apart from that it seemed a pretty typical scene.

Heading south on Wakefield Mound Road I passed a sign advertising 'Tony's House of Hope', which is seeking donations and will provide a women's facility and classroom. This is just past the United Steelworkers' office with its sign saying 'Unity and Strength for Workers'.

At Wakefield I was faced with a choice of route. Big Run Road is wide and heads towards the energy plant, whereas Salt Creek

Road is small and residential. I could go up either and then turn onto Rapp Hollow Road and come back on the other road. I decided to drive up Big Run Road and take the right turn along Rapp Hollow Road before turning sharp right onto Salt Creek Road, where I passed the Wakefield Freewill Baptist Church (Pastor Lowell King) with a goodly crowd of Dodges, Fords, Chryslers and Jeeps parked outside, suggesting that Pastor King had a substantial congregation for his 6:30 pm service. It was now nearly 8 pm.

Inside the six-mile loop that I had performed was the farmland where the last Passenger Pigeon in the wild had been shot. There were Mourning Doves flying over and Eastern Bluebirds nesting in boxes in gardens. On a sunny Sunday evening some families were letting Dad burn the meat on barbecues outside and other dads had donned baseball mitts and were throwing balls with their sons. Each house, whether well-kept or running a bit to seed, had a well-cut lawn and a mailbox on the roadside. There were basketball hoops on some house-ends and dogs on leashes in some yards. A few houses had an old metal water pump in the middle of the lawn.

The surroundings were wooded hills just like many other parts of southern Ohio, but also just like many other parts of the eastern USA. I hadn't expected anything different but just as ordinary-looking people can sometimes do amazing things, this ordinary-looking loop of American road contained the place where the Passenger Pigeon in effect was made extinct. The great abundance of the species ended right here in March 1900 as much as it ended in Cincinnati Zoo 14 years later. No Passenger Pigeon was reliably reported from the wild after Press Clay Southworth, aged 14, shot one in this ordinary patch of Ohio farmland and woodland.

That was my fourth homage of the day. Although we certainly know that Martha was the very last Passenger Pigeon on Earth it is much more difficult to be sure that Buttons was the last one in the wild. Perhaps similar scenes were played out in similar corners of the USA by similar boys with similar firearms, but if they were then they didn't occur much beyond 1900.

All that homage wrapped together in one day can leave one feeling hungry. The pleasant Asian receptionist at the motel had told me that Ritchie's did a good steak. But Ritchie's was shut, so it was take-away pizza from Gio's. I took the pizza back to my room, where, I was glad to hear, the couple in the next room had stopped making noisy love, and this made it easier to concentrate on writing and filing my July column for *Birdwatch* magazine – written in May, out in June, called July.

It had been a full Sunday. I had paid my respects to live birds in the early morning, and then to a dead Frenchman, a dead caged pigeon and a dead shot pigeon as the day wore on. Was it Buttons or Martha who really marked the end of the Passenger Pigeon as a species – or was it rather some unknown, unnamed pigeon back in time whose death marked the point at which extinction became inevitable? Was any book of bird paintings really worth $11.5m, and did the value of Audubon's Passenger Pigeon plate increase or decrease with the species' extinction?

20 May 2013

I woke early and felt completely rested. Every vestige of jet lag seemed to have dropped away. I headed back to see whether the area where Buttons was shot looked any different on a Monday morning from a Sunday evening.

In his research, Chris Cokinos discovered that Buttons had been shot by Press Clay Southworth on 24 March 1900 on the family farm. The bird had been stuffed and mounted by a Mrs Barnes, and was donated to the Ohio Historical Society by her husband, Mr Clay Barnes, Pike County's ex-sheriff, on 27 February 1915 – almost exactly six months after Martha's death.

Cokinos tracked down an account of the killing of Buttons by Press Clay Southworth himself, written in 1968 at the age of 82. He had been feeding cattle in the barn yard and saw a strange bird fly up and perch in a tree. As he was only 14 he had to persuade his mother to let him have the 12 gauge gun, and with it he brought down the strange pigeon from high in the tree. His mother identified it as a Passenger Pigeon because she remembered seeing them as a girl, as did his father later, and he took the bird to be stuffed.

This little tale summed up so much about the Passenger Pigeon. A country boy, presumably accustomed to seeing birds around the farm, saw an unfamiliar bird, and when he shot it and presented it to his parents they both recognised it from their childhood. In one generation this species had gone from familiar to unrecognised – and indeed to extinct in the wild when that shot was fired. And on that date in late March a few decades earlier there would have been millions of Passenger Pigeons courting, building nests and even perhaps laying the earliest eggs of the season.

As I drove around the loop road again I felt I wanted to knock on doors and tell people that where they lived was special in a rather sad way. There was no sign that the inhabitants of this area had the faintest idea of the role that their neighbourhood had played in the extinction of the commonest bird on Earth. No brown historical marker marked the scene, and there was no 'Passenger Pigeon Bar' to commemorate the event.

It was nearly 9 am. I waited in my car across the street for a few minutes for the Wakefield Grocery to open. Maybe I could ask in there about whether there was any lingering remembrance of the Passenger Pigeon locally. When I entered, a man was vacuuming the floor noisily and the woman behind the till didn't look in the mood for talking. I bought a Mound Bar and left without raising the subject – it didn't seem an opportune moment. This was Monday 20 May 2013, and Wakefield, Ohio, had forgotten the events of Tuesday 24 March 1900. In fact, the world had forgotten the Passenger Pigeon almost completely. And I guess that is too often what we mean by 'Progress' – forgetting the past rather than incorporating it in our present.

It was in that sombre mood that I started my search for a breakfast that would both fill me up and cheer me up – but some time later I actually had the worst breakfast I've ever had in the USA. So the names and locations in this tale have been changed to protect the guilty. But the food was not the worst part.

It started off promisingly with a friendly greeting from the man in charge (let's call him Gerald – not, definitely not, his real name). The place had a sort of lacklustre charm about it – or is that just my British politeness again? There were several men, who looked like regulars, already installed and I chatted to Gerald, who had spotted that I was not from these parts. I mentioned that I was interested in Passenger Pigeons and asked him whether he had heard of them. When he said he had, my hopes rose momentarily, but then he went on to tell me that there were lots of them each year at the county show and I nodded and smiled. I got coffee and ordered my usual eggs (over easy), home fries, as hash browns weren't available, a steak as a cheer-me-up treat and wheat toast.

The coffee wasn't good, and when the meal came the eggs and toast were OK but the steak and the home fries were poor. But the place was full of conversing locals so I ate my breakfast and listened to the conversation.

Gerald's regulars were all men above the age of, I guess, 60. Some came and some went but at any one time 80% of them had baseball caps on, and at any one time about a third of the baseball caps were emblazoned with the John Deere logo. We are talking rural Ohio here.

And that's important because rural Ohio is different from urban Ohio. In political terms, Ohio is a bit like my home constituency of Corby in Northamptonshire. It's a swing state, like Corby is a swing constituency. The voters of both change their minds when the country changes its mind.

Ohio is very good at it too – it rarely gets it wrong.

Most of Ohio, geographically, voted Republican in the last presidential election, but the cities were Democrat, and the three main cities are the three Cs: Columbus, Cleveland and Cincinnati. Because lots of people live in cities, the overall vote was narrowly for Barack Obama, the man of the 'Left'.

It's rather like that in Corby too – Corby votes Labour, the small towns vote one way or the other and the countryside votes Tory.

Where I had breakfast was rural and therefore could be expected to be Right-leaning in its politics.

The conversation was of the difficulties of growing potatoes and moved on to the subject of Raccoons as garden pests. Then one guy pipes up (let's call him Clarence) and asks:

'That Obama done good. Done reduced tax on us. What do you think of him now? He done good.'

This remark may have been uttered specifically to get the torrent of disagreement that it elicited – Clarence struck me as that type. One of Clarence's friends said:

'Don't trust him. Good at playing dumb.' (I thought about that all day; very profound.)

The conversation turned to agricultural matters for a while before Clarence said he had bought a couple of metal monkeys and they were rare ('Can't get 'em on the internet') but he was prepared to sell the pair, at profit, for $25. He talked about these two monkeys for a while, and the crowd got interested (and so was I, but trying not to show it), so Clarence got a monkey from the car.

It was an ugly little hollow cast thing. Clarence said it was made of 'that yellow metal' – someone said 'Gold?' and Clarence smiled. I don't think Clarence was anyone's fool.

There were offers of $2 on the table already but Clarence was holding out for more to recoup what he said he had paid for them. I don't think Clarence was any sort of fool.

There was apparently a little bit of yellow showing, and the conversation centred on them being brass monkeys.

One man said, 'It's dark.'

Clarence, 'Scratch and you'll see it's yellow.'

Other man, 'Scratch Obama – what colour'll he be?'

Clarence and others, 'Still black.'

Someone, 'Monkey here looks like Obama – might be his ancestor.'

General laughter, then someone asked for wheat toast and gravy.

I am quite shocked at the hatred for Obama in some parts of the USA. I have seen roadside signs saying 'US threatened by foreign leader – Obama!' and 'Obama bringing down US economy with spending policy.' Now the latter is a fair economic debating point but even I, as an ordinary citizen, haven't bothered to make a sign criticising George Osborne's economic policy and put it in my garden.

Is it just a political difference? Is it fuelled by the fact that Obama is young? Or good-looking? Maybe the fact that he is clever is the problem. Maybe it's all of these and more. Maybe, dare we say it, it is at least partly because he is black?

It was just an overheard conversation over a bad breakfast – but it left a worse taste in my mouth than did the tough steak that was the opposite of a treat.

I needed to move on and I wanted to see some old trees. My rather desultory research before leaving the UK had told me that the Wayne National Forest had some very old trees, so I parked outside the visitor centre at Nelsonville and asked where I might see the oldest trees they had. The staff on the desk were very helpful considering the vagueness and, I'm pretty sure about this, the unaccustomed nature of the request.

Sometimes one is very lucky. This day I was very lucky. Dawn McCarthy pointed me in the direction of her husband, Brian, who just happens to be a professor at Ohio University in Athens and is an expert on old-growth forest and vice-chair of the American Chestnut Foundation. Brian was kind enough to point me towards Dysart Woods, and a couple of hours later I was puffing a bit on the steep inclines of this ancient woodland.

Dysart Woods is a 20-hectare tract of old-growth oak forest and is the largest remaining fragment of the original forest of southeastern Ohio. I stood by oaks that were around 400 years old and more than 40 metres high with diameters of well over a metre. They were towering trees whose impact was slightly lessened by the fact that many of them clustered near the creek bed on the 'red' walking trail so that their great height was masked by growing at the bottom of the slope. However, this meant that as one walked easily down the slope, and less easily back up again, one could look at the full height and stature of the tall ancient trees. Some trees there have been dated to 600 years old – before Columbus arrived.

This forest may resemble the forests before Europeans invaded

America, with its well-spaced tall thick trees and a sparse
understorey of plants. The large trees were noticeable even to
someone like me who is not much interested in trees, and the
amount of dead wood scattered around was very different from a
typical, or even an exceptional, English woodland. There was lots
of rotting wood on the forest floor, including some huge tree
trunks, and there were dead trees standing in the forest that had
not yet toppled over.

When the Passenger Pigeon was still at its most abundant these
trees were alive. Some of them must, surely, have had Passenger
Pigeons perch in them 150, 250, 350 or more years ago. If they
could talk they would talk of the rivers of pigeons passing by in
search of acorns and beech mast. I put my arms as far around one
of the oaks as they would go – not far – and tried to feel vicariously
close to a Passenger Pigeon. It didn't really work on an emotional
level, but it made me think.

I dined at Pete's Place in Wheeling, just across the Ohio River
in West Virginia, although I hadn't intended to do so. I had had
breakfast there on a previous trip and as it had been a good breakfast
I had wanted to repeat it. But when I drove past, just to check, on
my way to find somewhere to stay the night, I noticed that it no
longer served breakfast on weekdays – and so I stopped for dinner.

Sitting out on the deck, looking over the Ohio River once more,
and watching the cottonwood seeds drift across the view, I ate my
steak salad (this steak was delicious) and drank gallons of iced Coca
Cola, and felt grateful that I had met Dawn, and that she had put
me in touch with Brian – otherwise I would not have seen Dysart
Woods and got a glimpse of how forests used to look.

I asked my waitress for advice on where to stay, somewhere
cheap, and she said that town was full because of all the oil people.
I didn't understand who all these oil people were – perhaps there
was a conference – but I did understand 'full'. I spent ages
wandering back westwards through Ohio and ended up 50-plus
miles back up the road in Cambridge – but even there it wasn't
straightforward to get a bed. The first place I saw, because it was
lit up in neon lights, was an 'American owned' motel but it was

full, because of all the oil people again. My guess is that 'American owned' is a way of signalling that the motel is not owned by someone of Asian origin, like Mr Patel in Henderson and the receptionist in Piketon, so I ended the day as I had started it, with a bit of modern-day American racism. However, the very nice American man did find me a room by phone and directed me to a dull but serviceable Holiday Inn Express where I fell quickly asleep while thinking of ancient forests and rivers of pigeons flooding the skies.

21 May 2013

Much of today was domestic (and much of what wasn't appears at the end of Chapter 6). I bought some water to keep hydrated as I drove, and filled up with gas at the same time. A woman at the gas station and I talked about the oil boom in this part of Ohio (everywhere has oilmen), which means that fuel prices are low (lots of competition) but room prices are high (lots of competition for a limited resource). She thought I had a nice Australian accent.

I spent quite a lot of time in Coshocton today. It was a convenient centre for things I wanted to do – see old trees and see a stuffed bird (tomorrow). But Coshocton was full of nice people – they may all have hated Obama, for all I knew, but they seemed very nice to me.

The woman in Walgreens, where I bought toothpaste and a new pair of cheap specs (having sat on one of my two pairs), pointed me in the direction of Bob Evans for breakfast, but nearby was a place much more my scene – Jerry's family restaurant – where I had two eggs over easy and hash browns (very nice too). Next door was the Country Squire Inn, where I stayed, and where the woman who checked me in knew nothing of the Passenger Pigeon.

A helpful woman in the optometrists on Main Street replaced a screw (no, I didn't have one loose) in my Ray-Bans and wouldn't take any money for it – not even a couple of dollars for charity. She hadn't heard of Passenger Pigeons but she might buy my book when I've written it. She also directed me to RadioShack, where I looked for a battery for my camera.

My plan was to take photos every day, but on my last day in the UK I realised the rechargeable battery was flat and wouldn't charge. I had been searching for a battery, not very hard so far, ever since. RadioShack seemed a possibility.

RadioShack didn't have the battery I wanted but they gave me some help. There was a man (see – I do talk to men too, when I must) and also a young woman. I asked about good places to eat and was pointed to The Warehouse (where, later, I had a very fine burger, salad and sweet potato fries, with Pepsi, for $8, because it was Tuesday and there was an offer on Tuesdays). I also bought a charger for my phone – see, very domestic. The young woman was very quiet and I said to her, 'You think I'm mad, don't you?' and she said, 'No, I wish you'd stay and talk to us all day.' I probably blushed, but since none of us had a working camera we'll never know.

Neither of them knew of the Passenger Pigeon but the guy confused it with the carrier pigeon – quite a few people do.

I went to see some more old trees. I drove north through the rolling Ohio countryside and through Amish country. Occasional horse-drawn buggies slowed down the traffic on the narrow roads. The women were wearing long dresses and their heads were covered, while the men had hats and beards. I remembered parties of Amish, recognisable by their clothing and because they were speaking German, being present at Magee Marsh, a bit north of here, two years ago during what Americans birders call 'The Big Week', when the place is full of migrating warblers. Despite their eschewing of modern technology I noticed that the birding Amish had top-of-the-range Swarovski binoculars like many of the other keen birders.

The Johnson Woods State Nature Preserve comprises about 80 hectares of old-growth deciduous woodland an hour and a bit north of Coshocton. It is a really lovely spot. The Baughman Township in Wayne County, Ohio, was used as a case study by Whitney and Somerlot to describe the changes in forest extent, composition and structure. In the early nineteenth century this township was still 90% forest – the main gaps were swamplands. The predominant trees were White Oak and Hickory on the coarser soils and American Beech and Sugar Maple on the well-drained soils, with American Elm and White Ash on the poorly drained soils.

By 1900, about 90% of the woodland in this township, and in much of Ohio as a whole, had been cleared for agriculture. The 9,000 hectares of Baughman Township's woodland had been reduced to around 900 hectares. The Johnson family acquired these woods in the 1820s and they protected the woodland whilst all around trees were being felled. This patch of the past deserves its local name of the Big Woods, as it is much larger than most other areas of old-growth forest that have survived in Ohio and the trees are impressively big.

There is lots of dead wood, and I heard lots of birdsong as I walked the boardwalk on this spring evening – as usual I wasn't sure what most of the song was, but a Red-eyed Vireo fed its young in a nest near the entrance. Some of the trees are enormous, mostly Red and White Oaks and Hickory, reaching to almost 40 metres with their first branches only appearing 15 metres above the ground. This is not a dark and forbidding forest but a light and airy one with the canopy high above one's head and the large trees spaced widely on the forest floor.

Many of the trees are more than 400 years old and were saplings before the Pilgrim Fathers (and Mothers) arrived at Plymouth Rock, Massachusetts, in 1620. It took another 200 years for large numbers of people to cover the 700-mile journey west to this part of Ohio, but from then on the impact on the landscape was rapid and dramatic.

This was another glimpse into what the woodland would have looked like when the Passenger Pigeon was the commonest bird on Earth. Wayne County was an area much used by the pigeons for winter roosts and was well within the breeding range too. Newman's Swamp, to the southeast of here, was a well-known and regularly used roost. Some of these trees must have had flocks of Passenger Pigeons perch on their branches in the centuries they lived before the bird's extinction. They would also have known Bison pass through the wood, for woodland Bison lived in these eastern woodlands into the end of the eighteenth century – and Grey Wolves and bears lasted much longer.

As I travelled back to my bed in Coshocton I mused on the fact that the countryside looked pretty similar to that around where I live in Northamptonshire – the size of the fields was similar and they were set in a similar wooded landscape. And each had precisely the same number of Passenger Pigeons – none at all – but in one case they were never present and in the other they have been lost. Nobody misses the Passenger Pigeon in Northamptonshire, but then they never knew it – and it seems that nobody misses it in Ohio either, because we have reached the time when it has slipped from memory.

22 May 2013

I liked Coshocton very much, but after an early breakfast in Jerry's (same meal as yesterday except with wheat toast instead of rye) I did some writing and then headed to the Ohio History Center in Columbus. I'd have gone sooner but it doesn't open until 10 am and it isn't open on Monday or Tuesday – so at 10:15 on Wednesday I arrived.

The satnav made the museum very easy to find and it was obvious when I was close as there were queues of those lovely yellow school buses arriving too. The big old-looking American school buses are as evocative of the USA as red telephone boxes and bobby's helmets are of the UK – except they are still a big part of modern American life. On the school run the buses stop regularly to pick up children waiting at the roadside with their books and satchels and a whole variety of expressions on their faces. The law is that you can't pass these buses while they are stopped and picking up which makes a lot of sense and says something about how America respects its children. So, when the school bus stops a 'stop' sign emerges from its side, its lights flash and everyone stops – even traffic on the other side of the road – and even when it's in the middle of the countryside with no other children in sight. I like that. These buses were already disgorging hordes of excited kids – so I fitted in well.

I paid my $10 entrance fee and asked the man on the desk whether he knew where I could find a Passenger Pigeon to look at. He thought it might be just down the aisle, but if not it would probably be in the natural history section. I thought this was a slightly offhand way to direct me to the remains of the last wild Passenger Pigeon known to man, but I did find Buttons in the natural history section in a case with other extinct species – Carolina Parakeet, Ivory-billed Woodpecker and Blue Pike.

Given her age and her manner of death I thought Buttons looked pretty perky really – although the unbiased might have said slightly dull and tatty. But this was the last wild specimen of the most abundant bird on Earth. Hardly anybody else gave her a glance. She suffered from there being a stuffed Bison just down the way: that got a lot of attention from the school parties that were now flooding through the ground floor of the museum, but none of the children stopped to look at a threadbare dead pigeon.

I thought back a couple of days to my drive around the area where Press Clay Southworth had shot this very bird. Rarely can one know so much about a stuffed specimen in a museum as I felt I did about Buttons. However, I was not the person who knew most or cared most about this stuffed female Passenger Pigeon, and nor was Chris Cokinos, who had stood looking at her in his time. When Buttons was housed in the Ohio State University Museum, some 50 years after he had shot her, Press Clay Southworth saw Buttons face to face again. Who knows what thoughts went through his head? There are few people who know that they banged in the last nail in the coffin of a species, and even fewer who can meet their victim face to face, albeit through the glass of a museum case, half a century later.

I stood for a while and looked at Buttons while excited young Ohioans rushed past. There is, though, a limit to how long you can look in awe at a stuffed bird, and my limit was about 10 minutes.

I had been in Kentucky and Ohio for almost a week, and I would pass through Ohio twice more on this journey, but now it was time to break free, and I had a long way to drive. First I needed to find a gas station as the 'low fuel' light had been blinking on the dashboard for quite a few miles.

Filling up with fuel in the USA can be an irritating experience. First, there is the business of working out whether you have to pay in advance inside or whether you can get your UK credit card to work at the pump, and then there is the business of spotting whether there is some lever you have to lift before you can get any fuel – but all that is only a little bit irritating compared with standing at the pump and realising that the fuel may contain up to 10% biofuel.

Whereas fossil fuels are long-dead plants and animals, biofuels are recently alive plants. They sound like a good idea in that if you use plants for fuel then you can grow them again and again and you may, basically, be recycling the same carbon and so your polluting impact is greatly lowered. That's the theory, but it doesn't work like that. First, there is a lot of energy which goes into growing and processing that plant material, which lessens the intended benefit considerably (sometimes completely). But secondly, if you are to use land to grow fuel it will take an awful lot of land. And it will take an awful lot of land that could be growing food. Since we aren't very good at eating less food, and there are more and more of us on the planet, land has to be found from somewhere to grow more food. That's a potent cause of rainforest destruction and ploughing up of grasslands across the world. And that habitat destruction is bad for wildlife, and also releases so much carbon that any advantage of biofuels on carbon grounds simply goes up in smoke. That's what really irritates me at the pump.

Refuelled, I headed west past Dayton, Ohio, where Orville Wright was born, and then about 60 miles later Millville, Indiana, where his elder brother Wilbur was born. At Indianapolis I turned northwest and headed through the flat Indiana countryside towards Chicago, Illinois, on I65. Nothing much happened on this journey, but the Benton County Windfarm in northern Indiana, and the nearby Fowler Ridge Windfarm, together have more than 300 turbines. Most of them were swirling around as I drove northwest, delivering 730 MW energy production, which is equivalent to about three-quarters of a typical UK nuclear reactor. I couldn't help wondering how a flock of a billion Passenger Pigeons would have coped with this windfarm.

I drove around Chicago with Ol' Blue Eyes on full volume singing 'My kind of town, Chicago is ...' over and over again as I crawled in the late-afternoon traffic. Still, it gave me time to view the Chicago skyline, where the Wrigley Building is still the most attractive tall building even though it is now dwarfed by many more modern piles.

In 1833, a couple of years before Chicago officially became a town, its inhabitants were shooting at passing flocks of Passenger Pigeons from their streets, from their roofs and from anywhere that would give them a clear view of the tasty birds heading northwards to their breeding grounds in Wisconsin and Michigan (and southern Canada). I too was heading to Wisconsin and Michigan, if only the traffic would clear.

I slept in Monroe, Wisconsin, the 'Swiss cheese capital of the USA', where there are cheese factories, cheese adverts and restaurants offering cheese curds everywhere. But my thoughts were of Buttons sitting behind the glass in the Ohio History Center. The late 1890s provided the last verified records of Passenger Pigeons for most US states. Did those last birds mope around the countryside as lonely individuals wondering where were the missing millions and billions? What was the true end of the line? Were there still a few more individuals into the early years of the twentieth century? I expect there were, but it hardly matters as they were by then, it seems, doomed.

And it seemed to me, although I admit I have a special interest in this subject, that the Ohio Historical Society should be making more of Buttons – she should, even at this late stage, be pampered and celebrated. She was the last one in the wild and her story is well documented. Surely on 24 March each year there should be some marking of the anniversary of the death of Buttons – and surely on 1 September 2014 there should be some public event to mark the centenary of the passing of Martha?

23 May 2013
I drove to Wyalusing State Park because it has a monument to the Passenger Pigeon – they say the first monument to an extinct bird in the world.

When I arrived, Bev, the ranger at the entrance, was talking to two couples. She was telling them where they might find Cerulean Warbler, Yellow-throated Warbler and Henslow's Sparrow. But she was also plugging the Passenger Pigeon Memorial hard. In the reception area there was a stuffed male Passenger Pigeon and three different pictures of the bird too.

I waited patiently for my turn, which wasn't difficult as I looked at the pigeon stuff and the Ruby-throated Hummingbirds on the feeder, and then said 'I've come to see the Passenger Pigeon Memorial, but I'd like to see some live birds too, so can you go over all that again, please?' A big, really big, smile spread over Bev's face, and we started talking Passenger Pigeons. A kindred spirit at last. Bev – where have you been all my life?

Should I stop and look and listen for Henslow's Sparrow before visiting the memorial? No, I don't think so. Life's sometimes too short to bother with sparrows.

Should I stop and look/listen for Yellow-throated Warblers, as their spot was right by my route to the memorial? No, they'll probably be there when I come back.

So, straight to the memorial – a simple metal plate, set in stone, in a wonderfully beautiful setting. It's high on a wooded ridge overlooking the confluence of the Mississippi and Wisconsin rivers. This morning, having it to myself, it was an idyllic spot. The sun came out, an immature Bald Eagle circled below me and then, gaining height, above me, and a female Rose-breasted Grosbeak was building a nest.

The words on the plaque are very simple:

Dedicated to the last Wisconsin Passenger Pigeon shot at Babcock, Sept 1899.

This species became extinct through the avarice and thoughtlessness of Man.

Erected by the Wisconsin Society for Ornithology

It's a beautiful location, and an appropriate tribute (and admonition). I was quite moved and I was glad I had come, and glad that I had the place to myself for the time I was there.

An account of the shooting of that last Wisconsin Passenger Pigeon at Babcock was published in the shooting magazine *Forest and Stream* in September 1899. The editor of that magazine, Emerson Hough, had been present, and he wrote:

> While we were cleaning our birds at lunch time on the first day, our guide Varney pulled out of his pocket some Turtle Doves [Mourning Doves] which he had innocently been shooting that morning. Among these was a bird to which he called our attention, saying it was 'too big for a dove' and he did not know what it was. 'Why, that's a pigeon!' cried Mr. Brown. 'It's a young wild pigeon.' And so it proved. The bird was about two-thirds grown and the plumage was yet pale and devoid of the fine luster of the adult bird. The tail feathers were pulled out in the pocket of Varney's hunting coat, but I got them and have them now, with the skin of the bird, which I secured. I cannot give many details regarding the killing of this bird, except that it was shot from a tree early in the morning by Varney. There were a lot of doves hanging around a buckwheat field and some of these lighted on a tree. Varney fired at the largest bird he saw on the tree, and put it in his pocket, thinking that it was a dove. It was nearly twice as large and heavy as a dove when we came to place the two birds together. Mr. Brown tells me that he and his wife have seen these birds in northern Wisconsin within six years, and they were once abundant all over this country where we were hunting. The *Forest and Stream* has always been very anxious to secure any positive proof of the appearance of this wild pigeon, and here is proof which is direct and unmistakable. It was the last feature needed to make my little Babcock experience a curious and enjoyable one.

It seems that the last Wisconsin Passenger Pigeon was killed with the same lack of appreciation of the importance of the event as was Buttons, six months later and 650 miles southeast in Ohio.

One of the people who attended the original dedication of the memorial on 11 May 1947 was the American ecologist and writer Aldo Leopold, in the year before he died aged 61. Leopold wrote of the event as follows in his essay 'On a monument to the pigeon':

> Men still live who, in their youth, remember pigeons. Trees still live who, in their youth, were shaken by a living wind. But a decade hence only the oldest oaks will remember, and at long last only the hills will know.
>
> There will always be pigeons in books and in museums, but these are effigies and images, dead to all hardships and to all delights.
>
> Book-pigeons cannot dive out of a cloud to make the deer run for cover, or clap their wings in thunderous applause of mast-laden woods. Book-pigeons cannot breakfast on new-mown wheat in Minnesota, and dine on blueberries in Canada. They know no urge of seasons; they feel no kiss of sun, no lash of wind and weather. They live forever by not living at all.

I've only recently discovered Leopold's writing, although I've known the name for ages. His most famous work is the *Sand County Almanac*, set on his farm here in Wisconsin. It's a series of essays (you might even describe them as the forerunners of blogs), and the passage on the Passenger Pigeon comes from that book – it's a lovely read. Leopold's account of Upland Sandpipers is also wonderful, as are his musings on conservation ethics, and we should all try to follow his thoughts about thinking like a mountain.

Leopold's daughter and others gathered on 1 June 1997 at the Passenger Pigeon memorial to mark its 50th anniversary. I imagine there will be gatherings in 2014 too to mark the passing of Martha. I hope so, as it is a delightful spot with a glorious view – a good place to remember lost beauty and to celebrate what we have left.

But there is only so much time that one can spend mourning the past – maybe that's our problem – and so I left the Passenger Pigeon Memorial, although my thoughts will often take me back. In a few minutes I was listening to a Yellow-throated Warbler singing from the top of a pine tree, and in the same tree I saw a Yellow-billed Cuckoo – two lifers together.

I went to the spot where a couple of days earlier 30 Cerulean Warblers had been seen, but with little hope. That many meant they were on migration, and in spring birds don't stop long. There had been no sightings yesterday so I guessed they had gone. I skipped the Henslow's Sparrow on the way out too, but I checked in with Bev to tell her how much I had enjoyed the Passenger Pigeon Memorial. It's just a plaque in a stone on a hill – but it meant a lot to me.

Sand County was really Sauk County, and it wasn't that far away, and it was in the direction I was heading, so I decided to visit the Leopold Center which sits in the woods and commemorates Leopold's life.

I like Wisconsin. It's very rural. My drive, of 90 minutes or so, followed the Wisconsin River through wooded hills and open cultivated valleys. I saw more Bald Eagles, Great Egrets and Sandhill Cranes. As the scenery passed it reminded me of Scotland; somewhere a bit like the Grampian farmland with the Dee or the Don passing through it. There was little traffic, there were few people, and it was a lovely drive.

As I approached the Leopold Center I had to slow down and go slightly off the road to avoid running over a really big terrapin. I had seen quite a few on the roads, mostly tiny ones, and some squashed ones, and I didn't want to arrive at the Leopold Center with squashed terrapin on my tyres.

I had a quick look at the exhibition and interpretation, chatted to the young woman on reception, bought some postcards of the cabin that Leopold built amongst the fall colours and chose not to spend $7 looking at the cabin itself. It was enough to look around the very well-designed exhibition and be reminded of some of the lines he wrote. Again, it was a kind of homage.

But the day had yet more to offer. It had been pastoral and rural and contemplative so far, but I knew that was about to change. Before it did, I drove carefully back along the road where the terrapin had been, hoping not to find it squashed by another, but it had passed safely into the wet woods. A few minutes later I was passing the Ho-Chunk Casino which offered me table games, bingo, slot machines and off-track betting. Aldo Leopold would have been so proud.

Only 15 minutes' drive away from the Leopold Center, through the quiet Wisconsin countryside, is Wisconsin Dells – the tourist trap to end all tourist traps! Water parks, theme parks, casinos and everything kitsch under the sun is here. A mixture of Blackpool and Alton Towers with America thrown in. It didn't appeal to me, but I don't want to be snooty about it. After all, I had spent the day so far visiting a monument to a dead bird and, really, another to a dead man – I'm the weird one.

You can take a tour on an amphibious truck or see, as I did from the road, an amazing upside-down White House called Top Secret Inc. – where you may learn, and it may not come as a surprise, that most recent US presidents were built by robots. The Kalahari Resort has an African theme (really?) and is the largest indoor water park in the Dells under one roof. There are lots of others too.

I wanted to see Wisconsin Dells because it has a past as well as a present, and its past is part of the Passenger Pigeon story. Wisconsin Dells was once Kilbourn City, the site of the 1871 colony which Schorger and I estimated at 136 million and 176 million birds, respectively. Perhaps close to the whole of the already much-reduced Passenger Pigeon population nested there that year. The colony was divided into two arms, one stretching from Kilbourn City northwest to Black River Falls and the other northeast almost to Wisconsin Falls.

For those not familiar with the geography of Wisconsin, Black River Falls is 80 miles away and Wisconsin Falls 50 miles away, and the total area covered by the colony was about 850 square miles (2,200 square kilometres), with some trees holding up to 100 nests. Imagine it! Can you imagine it? I wasn't sure that I could, which is why I came to experience it – I wanted to see the ground.

Of course there would have been gaps in the colony, and there may have been some exaggeration (but there may have been some hesitancy at seemingly telling a tall story too). But the mind can't really take in the number of birds that might have been involved.

The juxtaposition of the modern Wisconsin Dells and what was

probably the largest described Passenger Pigeon nesting –
overlapping in space but separated by 140 years – sums up rather a
lot about 'Progress', to my mind.

I was heading northeast, so I drove up the length of the
shorter arm of the colony – just 50 miles. After passing
Wisconsin Dolls Gentlemen's Club, I tried to imagine what the
landscape would have been like full of nesting pigeons. I looked
across to a range of hills running parallel to the river and tried
to imagine those woods as part of an enormous colony of
pigeons. I looked at a wooded hill in Friendship (25 miles out of
Wisconsin Dells) and tried to picture it covered with Passenger
Pigeons. It's very difficult, even for me, and I'm hooked, to keep
thinking Passenger Pigeon for a 50-mile drive through the
countryside – but 140 years ago this site was covered with
pigeons. In 1871 there were well over 100 million of them in
this area and yet, just 28 years later, the last Wisconsin bird was
killed. And 29 years later the last wild one on Earth was shot.

If there had been 100 Passenger Pigeon nests in one tree that I
passed, or a single hectare of packed colony, then I could have
pictured what was missing, but because everything was missing it
was very hard.

I spent the night in the Hiawatha Motel in Escanaba, Michigan.
Michigan was the 11th state I had entered since arriving in the USA
a week earlier: Texas, Louisiana, Arkansas and Missouri were on
my journey north, and then I had been a pigeon explorer in
Kentucky, Ohio, West Virginia, Indiana, Illinois, Wisconsin and
Michigan.

I was just skimming the surface. I had spent a single day in
Wisconsin but it had given me a glance at the Passenger Pigeon
Memorial and a glance at Bev's smile as she recognised a kindred
spirit. And then, fittingly, I had been able to visit where Aldo
Leopold lived in the woods, and where he wrote of Passenger
Pigeons and nature, before seeing the present-day Wisconsin Dells
and trying to imagine what it might have been like when it was
called Kilbourn City. People flock to Wisconsin Dells, but would
they flock to see one of the wonders of the natural world if the

pigeons were still around? I wonder. I would – but then I'm a bit odd – but I bet Bev would come too.

24 May 2013

It was Bob Dylan's birthday so I started the day with *Blood on the Tracks*, and as I listened to the first track, *Tangled up in Blue*, it seemed appropriate that early that morning the sun was shining and it shone all day long as I drove through Michigan.

I drove with Lake Michigan on my right. It's like being by the seaside – they're not called the Great Lakes for nothing.

I almost gave up glancing at the lake to see if there were any birds because there didn't seem to be any, and then I saw two Great Northern Divers (Common Loons, if you will). At a brief stop I confirmed that the gulls I kept seeing were indeed Ring-billed and I saw a Blackpoll Warbler. A Raven or two passed overhead. I drove over the Mackinac Bridge between St Ignace (on the north side) and Mackinaw City (on the south), which crosses the strait where Lake Michigan joins Lake Huron. I was heading to Petoskey, a city with a big role in the Passenger Pigeon decline. The name Petoskey comes, it seems, from the Odawa Indian and means 'where the light shines through the clouds'; in the late 1870s there were clouds of Passenger Pigeons here.

I got to Petoskey around noon and, by chance, by happy happenchance, I spotted a museum.

The Little Traverse Historical Society Museum had information about Passenger Pigeons and a very attractive large painting, at the end of the room, of folk collecting pigeon squabs in the woods as they did in prodigious numbers here in 1878.

Passenger Pigeons were numerous in this area in 1876. The local paper, the *Emmett County Democrat*, carried its first story about the abundance that year of Passenger Pigeons on 26 May ('Pigeon shooting is the order of the day') and followed up with stories through June of the sounds of gunfire resembling the celebrations of July 4th and of over-excited, or simply wildly inaccurate, pigeon shooters wounding horses with their shots. Most of the shooting around Petoskey was probably done by sportsmen rather than for commerce. However, 30 or so miles away at Cheboygan, on the Lake Huron side of the Michigan peninsula, more than 200,000 Passenger Pigeons were shipped out by steamer.

Fewer birds were seen in 1877, but on 29 March 1878 the paper carried the following: 'Great flocks of pigeons are seen in all directions recently, and almost every man and boy is seen with musket walking towards the woods.' Three men were said to have captured 400 Passenger Pigeons in a day, and a week later the *Democrat* stated that there were 'lots of bird hunters in town'.

There were about 2,000 people associated with the pigeon-killing – pigeon catchers, pigeon pluckers, barrel makers and ice suppliers all had a bonanza. Emmet County, which occupied the whole area of my drive from the Mackinac Bridge to Petoskey, a distance of 46 miles, was formed in 1840. The 1860 and 1870 censuses for the county enumerate just over 1,000 people, and yet, with the coming of the railroad in 1874 and the coming of the Passenger Pigeon too, the 1880 census showed a five-fold increase in population to more than 6,600 souls. The Passenger Pigeon business boosted Petoskey's economy and prospects. Many must have thought that they had a nice little earner for life here.

During the first week in April, 18,000 pigeons were shipped from Petoskey every day, and this was just the start of the 1878 season. A little later in the month the figure was around 45,000 pigeons a day. Squabs were poked out of nests with sticks and shot out of nests with arrows; fires were lit under trees and sulphur burned in order to make the young pigeons jump from their nests; adults were shot and netted by a variety of ingenious means using decoy birds and salt licks to attract the victims. It was carnage. Petoskey was given over to pigeon-killing.

New nestings were discovered, perhaps re-nesting attempts of birds whose squabs had been plundered as well as some new birds arriving, so that the season lasted into late May, by which time at least 1.1 million Passenger Pigeons were despatched from Petoskey and nearby Boyne Falls and Cheboygan by rail and boat. Around four-fifths of the birds were dead, but around a quarter of a million birds were transported alive. A traveller on a steamer talked of the birds being packed in crates from which they made a terrible noise, preventing the human travellers from sleeping.

The number of birds killed in this colony must have been higher than those sent off for food. There will have been young that died because their parents were killed, birds killed but not recovered, birds scared away by the proceedings whose squabs starved, birds eaten or preserved locally, and any manner of other losses along the way. Some say that as many as five million Passenger Pigeons lost their lives in this nesting.

But it was a huge colony – 40 miles long and a few miles wide. Who knows how many pigeons were there to start with?

All looked forward to the pigeons coming back next year, or soon after, but they never came back in similar numbers. In 1879 it seems that the Passenger Pigeons nested further north, but hopes were high for 1880 because the beech mast crop was good. The *Democrat* wrote, 'As there is a large crop of beechnuts this year, it is fair to presume that next spring will see as many pigeons in this part of the country, and perhaps more than there were last summer and our friends in the rural districts are making preparations of catching and shipping them.' But that was wishful thinking. There were still 40 barrels shipped from Petoskey on 18 May, but the frenzied slaughter of 1878 was never repeated. 1881 seemed to be a year when many pigeons passed through but nested further north and 1882 was a dismal failure for pigeon hunters. And both the Passenger Pigeons and the pigeon hunters were soon gone completely.

The telegraph and the railroad made it all possible, and the fact that Petoskey had boat links was important too. Other places, and this place at earlier times, could not have turned pigeon-killing into such a profitable industry for thousands of people – the local paper described the seven-week period of the 'big slaughter' as having put into circulation locally between $30,000 and $40,000.

I had a long chat with Michael Federspiel at the museum, and I am grateful to him for his kindness and his help. He pointed me in the direction of another Passenger Pigeon memorial a few miles back up the road I had travelled.

As I drove, I tried to picture the young Ernest Hemingway spending his vacations in this area – fishing in these lakes and hunting in these woods. The woods around Walloon Lake, just south of Petoskey and still a place of holiday homes, were where Hemingway learned of the old-growth woodland – which was already mostly gone by the time the great slaughter occurred and was probably completely absent 20 years later when Hemingway explored the area as a child in the first and second decades of the twentieth century.

Hemingway wrote of the virgin forest in his Nick Adams stories, and most particularly in *The Last Good Country*, in which the eponymous hero and his sister hide, as youths, in the forest, where:

> there was no underbrush and the trunks of the trees rose sixty feet high before there were any branches. It was cool in the shade of the trees and high up in them Nick could hear the breeze that was rising. No sun came through as they walked and Nick knew there would be no sun through the high top branches until nearly noon.

The siblings admit to each other that these woods making them feel strange and Nick says they make him feel 'the way I ought to feel in church' and that 'This is the way forests were in the olden days. This is about the last good country there is left.'

Nick's sister says 'this kind of woods makes me feel awfully religious.' And Nick replies 'That's why they build cathedrals to be like this.'

She asks whether he has ever seen a cathedral and he says, 'No. But I've read about them and I can imagine them. This is the best one we have around here.'

Hemingway probably never saw such forests, because they had been cut down, but he would have heard of their recent existence from local people, just as he heard of the Native Americans of whom he also wrote. We can be quite sure that he also heard of the Passenger Pigeon slaughter.

The historical marker is at Oden (by the fish hatchery, but the railway carriages by the side of the road are the best landmark). The front of the memorial has an image of a Passenger Pigeon and some general information. The back has:

> At one time Michigan was a favorite nesting ground for the Passenger Pigeons. Vast quantities of beechnuts and other food attracted them. Each spring immense flocks arrived, literally darkening the skies hours at a time as they flew over. Here at Crooked Lake a nesting in 1878 covered 90 square miles. Millions of birds were killed, packed in barrels and shipped from Petoskey. Such wanton slaughter helped to make the Passenger Pigeon extinct by 1914. The conservationist's voice was heard too late.

I'm tempted to say, in that gloomy way that conservationists do, 'twas ever thus' – but tomorrow I hope to see a conservation success story.

25 May 2013
The Mio Motel, in Mio, Michigan, hasn't got uniformly great write-ups on *Trip Advisor* but I liked it. The guy in charge is a bit brusque, but by the standards of British customer care he is in the 'eccentric and a bit curt' category, no worse.

The big advantage of the Mio Motel is that it is only 400 metres from the Forest Service office where, if you are there at 6:45 am at the right time of year (and pay $10) they will introduce you to one of the biggest conservation successes on Earth and show you one of the world's rarest birds – Kirtland's Warbler.

Seven of us were there today and we heard, from Tim, about this species' need for young Jack Pines – 5–20 years old. It's a fussy bird, rather like the Woodlark and Nightjar back home. A difference is that Kirtland's Warbler only lives in this part of Michigan, in a small part of Canada, and in Wisconsin, over the other side of Lake Michigan (where there are fewer than five pairs).

In 1951 there were 500 pairs, and a similar number were counted in 1961. But in 1971 there were just under 200 pairs – in the world. Conservationists, foresters and birders leapt into action, and now there are more than 2,000 pairs. Still not a huge number, is it?

The successful recipe was to create large areas of the right-aged trees (by clear-felling) and to bump off 3,000–4,000 Brown-headed Cowbirds a year.

Clear-felling replaced the role of wildfires in maintaining enough large areas of the right sort of habitat for this picky bird (which winters in the Bahamas – lucky thing!). The cowbirds are nest parasites (like cuckoos) and were Great Plains birds until we cut down forests to create farmland, and then they moved in. Kirtland's Warblers seem pretty susceptible – 70% of nests were affected by cowbird eggs and nestlings before control, but only 6% after.

After the excellent briefing we drove to a Kirtland's Warbler patch. As soon as we parked in a place which could easily have been in Suffolk or Dorset (sandy soil, pine trees of different ages in blocks) we heard a distant Kirtland's Warbler. During a short walk we heard lots more Kirtland's Warblers and saw several too – males sitting at or near the tops of trees and singing away. They are proper American warblers – yellow and black and well-marked.

By 9 am I was heading south after a very enjoyable and successful visit. There's quite a contrast between the fate of the once superabundant Passenger Pigeon and the always quite rare Kirtland's Warbler.

Kirtland's Warbler may come off the Endangered Species Act some time. I hope that doesn't mean that the money disappears – otherwise its numbers will plummet again. No doubt Jared Potter Kirtland, after whom this cracking warbler was named (as was a cute slug-eating snake) would have known the Passenger Pigeon well. He lived from 1793 until 1877 and was born in Connecticut but lived most of his life in Ohio.

Kirtland was a naturalist, and if he had been asked which of the two species, Passenger Pigeon or Kirtland's Warbler, would be the first to be driven to extinction then surely, for most of his years at least, there would have been no question in his mind that his warbler was the more vulnerable, by far. Kirtland's Warbler was always a species of restricted range, whereas Kirtland must have seen the sky darken with Passenger Pigeons often through his life. But maybe in his old age he would have read of the growing toll on the Passenger Pigeon from shooting for the eastern and southern markets, and he may even have noticed that the flocks he saw seemed to come round less frequently, or that they were a little smaller than they had been in his youth, and he perhaps wondered whether his memory was letting him down and exaggerating the previous abundance of the bird.

When Kirtland's Warbler dipped to below 200 pairs there was still plenty of time to catch its falling population before it hit the ground. The Passenger Pigeon, however, seems to have been doomed when it was reduced to its last few millions of birds. Why didn't the Passenger Pigeon hang on, maybe in very small numbers, for longer?

I puzzled over this with little profit as I headed south past Detroit and through Ann Arbor. I crossed back into Ohio near Toledo and just had to stop at Magee Marsh, even though it was late in the day and late in the season, to remind myself of the fantastic views of warblers that I had had there two years and a few weeks earlier. I saw Chestnut-sided, Magnolia and Yellow Warblers and Common Yellowthroat, as well as Red-breasted Nuthatch and Philadelphia Vireo, so it certainly wasn't a wasted stop.

I also stopped at a roadside ice-cream bar and made the mistake of ordering a medium-sized chocolate ice-cream. It turned out to be only slightly smaller than my car, and eating it delayed my onward journey.

As I drove east I thought about how Kirtland's Warbler is somewhat similar to our Dartford Warbler. It lives in early successional stages or scrubby vegetation with scattered trees – although the Jack Pines didn't have a Thomas Hardy to sing their praises in quite the same way as he did for Egdon Heath in his Wessex novels. And that got me thinking about climate change, as the Dartford Warbler is a species which will be, and is being, greatly affected by a warming climate caused by people flying across the Atlantic Ocean and driving around the USA's roads (and by a couple of centuries of fossil fuel use too, of course).

Dartford Warblers are increasing in numbers in the UK, and have done ever since the population was reduced to fewer than a dozen birds by the snows of the 1962/63 winter – which I remember, as a four-year-old, as being very good for snowmen. Since those days, Dartford Warblers have increased to thousands of pairs and now nest in counties as widespread and dispersed from their Dorset and Hampshire refuges as Cornwall and Suffolk. In the UK they are spreading north all the time in response to climate change. They have already reached north Wales, and we should expect to see them nesting in the Peak District and the North York Moors in a few years. But the rub is that their Iberian populations, which are much larger than the UK's current population and much larger than I can imagine the UK population being, are declining, also in response to climate change. The species is shifting its distribution north – which looks like a bold expansion at the north of the species' range but a craven retreat when you look at the southern edge of the range. In the last three days I had driven through most of Kirtland's Warbler's world range, and it occurred to me that if it shifts north too then someone ought to tell the Canadians that it is coming and that they ought to be preparing the ground, literally, for it. There are Jack Pines in Canada, but Kirtland's are fussy warblers and they need large patches of suitable habitat in order to thrive. I hope Canada is preparing the welcome mat for this fussy little warbler.

And, I wonder, how might climate change have affected the Passenger Pigeon if it still existed?

I was determined to get beyond Cleveland this evening (you have to have goals in life), partly so that I could have an easy day's driving the next day when I needed to arrive in upper New York State at Ithaca. While travelling in the USA I have quite often had the experience of passing several motels and thinking 'I'll stop at the next one if it looks OK' and then driving for miles and miles because there seem to be no places to lay one's head. That had happened to me two nights earlier in Wisconsin where I had started looking for somewhere to stay north of Green Bay. I pinned my hopes on Marinette on the Michigan border, but ended up staying more than 50 miles up the road in Esconaba, Michigan.

The same happened this evening as I drifted east through Ohio and just into western Pennsylvania. I was tempted to call in at the first *River Rock At The Amp* concert of the summer – eastern Ohio's answer to Glastonbury – but I kept going, past huge numbers of motorbikes, and then wasn't surprised to learn that a Led Zeppelin tribute band, ZOSO, were heading things up. When I found the Royal Motel in Hermitage, Pennsylvania, I was lucky to get the last room.

Driving 400 miles today was my contribution to pushing Kirtland's Warbler north through climate change, but I had had first-hand experience of one of the most uplifting conservation success stories of which I know.

26 May 2013

Pennsylvania isn't named after the Penns for nothing – and nor is it called *sylvania* for nothing. It's full of trees. It's the most tree-rich state I have seen.

As I drove up the Allegheny River's course there were trees everywhere. Trees and rivers. It was May, so the mayflies were flying. Here in the USA the mayflies are bigger than ours – as you might expect. The Allegheny had its fishermen, and the occasional fisherwoman, trying to fool the fish into taking their flies rather than the juicy large mayflies on offer. I have no idea if the fish were fooled or not.

This was once the land of the Seneca, an Iroquois people. Once a white Passenger Pigeon had come to an old man and told him that the Passenger Pigeon had been selected, after a discussion among all the birds, to be one that gave man a tribute of food.

Horatio Jones, a European invader captured as a teenager by the Seneca, told of a day when they were camped by this same Allegheny River, when a runner arrived and said there were pigeons nesting at two days' distance. The whole village upped sticks and made the journey, meeting other tribes converging on the pigeons from other directions. Jones wrote that 'it was a festival season and even the meanest dog in the camp had its fill of pigeon meat.'

I spent the day in a leisurely way driving through wooded valleys and over wooded hills to Ithaca, New York, where I would be staying with friends and visiting the world-famous Cornell Lab of Ornithology.

My friends Chris Wood and Jessie Barry live in the woods, and the decking where we sat talking birds over dinner provided Common Yellowthroat, Northern Cardinal, Hairy and Red-bellied Woodpeckers, Rose-breasted Grosbeak, Scarlet Tanager, Gray Catbird, Ruby-throated Hummingbird and more. An earlier short walk with Jessie had added Swamp Sparrow, House Wren, Pine and Blue-winged Warblers, Eastern Towhee, Black-capped Chickadee, Veery and Alder Flycatcher to the trip list – it's good to have friends who are birders!

27 May 2013

It was a holiday in the USA, just as it was back in the UK. Here it was Memorial Day, when those who fell in battle are remembered. There were Stars and Stripes all over – and I like that. Remembrance Day in the UK, with its poppies, is moving, but the only time you see anything like the number of flags back home is when England are just about to lose in some international football tournament.

I spent the day birding with Chris Wood and Matt Young, another friend from Cornell. It was so good to be with people who knew all the calls and songs as we travelled around Cayuga County. We got 120 species in a pretty leisurely birding tour – and had scones (which weren't scones but were very tasty, and were sold in a store to the sound of Bruce Springsteen).

We talked as we drove – about birding, about the challenges facing nature conservation, about people, about birds, about my accent, about American spelling, about wine, about whether England is to the European Union as Texas is to the USA, about state gas and car taxes, about radio programmes. We talked about a ton of things, but it always came back to birds.

The birding itself began. We nibbled on Caspian Tern and Least Sandpiper before a starter of Bobolink (with Savannah Sparrow accompaniment), followed by a main course of Cerulean Warbler with a side of Mourning Warbler. Dessert was a very tasty Upland Sandpiper, with a cheese course of Dunlin, Shortbilled Dowitcher, Semipalmated Sandpiper and Plover, White-rumped Sandpiper and more. The mint with the coffee was a Bonaparte's Gull. It was quite a feast.

Cerulean Warbler was a long-awaited delicacy. We had great views of a difficult but gorgeous bird. This is a species that hugs the treetops, but we saw one out in the open and very well. There it was in its cerulean blueness – stunning. It sat out on a bare branch and slightly cocked its tail and drooped its wings and we feasted on its beauty. There is not a snowball's chance in Hades that I would have found this bird on my own.

But, and it seems slightly disloyal to the Cerulean Warbler to say this, I think the Bobolinks may stay more firmly in my mind for longer. Gorgeous, but declining, birds of grassland, the males were everywhere at one meadow, singing and fluttering like Skylarks across their grassland home. This was a glimpse into what meadow bird densities were like before silage and early cutting caused bird declines.

If the Passenger Pigeon were still alive and kicking, would we have headed off to its nearest enormous colony to see the sight of millions of birds coming and going? Would we have seen skeins of birds racing across the New York skies and along the shores of Lake Cayuga? We might have done, as the Montezuma saltmarshes, where we saw flocks of shorebirds, Black Duck, Alder Flycatcher and Marsh Wren, were favoured feeding places for the Passenger Pigeon. What would that day have been like?

A little over an hour's drive further east would have taken us to Utica, where during the May 1871 Annual Convention of the New York State Sportsmen's Association 10,800 live Passenger Pigeons were ordered for targets – probably from the colony whose location I had visited at Wisconsin Dells a few days earlier. Hundreds of thousands of birds from this colony would also have been used as targets at shooting competitions such as those in Toledo, Ohio, Buffalo, New York, Milwaukee, Michigan and Washington DC.

The living birds we saw in May 2013 will stick in the mind for a long time, but the companionship of friends will stay as a memory even longer. That evening, in a fabulous meal at the home of John Fitzpatrick, who is the leader of the Cornell Lab, I tasted wild rice for the first time – it's very good. Some thought that Passenger Pigeons had cultivated rice in their guts when shot in northern states far from where rice was grown, but Schorger believed that they had mistaken wild rice for the cultivar.

28 May 2013
I had meetings with quite a few people at the Cornell Lab in Sapsucker Woods and then gave a talk in the evening. I was making new friends all the time and strengthening the links with existing ones. Everybody was very kind and helpful.

Cornell is linked with another of the USA's lost birds – or is it an almost-lost bird? On 3 June 2005 the Cornell Lab published a scientific paper in the journal *Science* entitled 'Ivory-billed Woodpecker (*Campephilus principalis*) persists in continental North America', which set out the evidence for there being a male Ivory-billed Woodpecker living in the Big Woods region of eastern Arkansas. The evidence consists of some observations and some rather fuzzy video footage which has allowed speculation and debate to rage over the years. Some people believe the record, others definitely do not, and many are undecided but hope that there are still a few individuals of this magnificent beast out there in the woods of the southeastern USA.

Fitz gave me a copy of the paper and I asked him to sign it, which he did with the words, 'For Mark, hoping one or the other of us sees one some day!' Fitz, although senior author on the paper, and the main recipient of the scepticism and derision, did not see the bird himself.

I read the paper at the time it was published and have read it again a couple of times, and it is suggestive of Ivory-billed

Woodpecker but not strong enough evidence to be totally convincing. But, as with any good piece of science, it lays out the evidence, draws some conclusions, and allows others to draw their conclusions too. As time passes, hopes fade for someone to produce definitive film footage of this bird.

Possible sightings of Ivory-bills still come in to Cornell. Maybe one day someone will get a perfect image of a perfect Ivory-bill and the world can rejoice, although it seems quite likely that if any still exist they are so few, and so far apart, that they are, in Fitz's words, 'ecologically extinct' even if there is the odd 'Martha' still out there. I guess we'll see.

The Ivory-billed Woodpecker is on a list, with Bachman's Warbler and Eskimo Curlew, of American birds that appear to have slipped into extinction but where there is still a glimmer of hope that they may be rediscovered. The Passenger Pigeon is on a different list, with Carolina Parakeet, Labrador Duck, Heath Hen and Great Auk, where all hope has gone. Or has it? We talked at Cornell of the idea mooted by some that the Passenger Pigeon would be a candidate species for reconstituting from the DNA preserved in museum specimens. It sounds pretty far-fetched to me, but it raises the question of whether we would really want to see the Passenger Pigeon back in our skies again.

Tom Schulenberg showed me some of the skins in the Cornell collection. Pulling out a large drawer he revealed Bachman's Warblers, Eskimo Curlew and Carolina Parakeets. And I held a Passenger Pigeon in my hands for a while. It was a male that had been shot nearby in the 1890s – and was in very fine condition except for having breathed its last breath more than a century ago.

This Passenger Pigeon was a strong-looking bird with a long graduated tail and an iridescent sheen on the side of the neck. My hands shook as I held it – I'm happy to admit it – it was a moving moment. As I handed the bird back to Tom he told me that I should wash my hands, as some of these old specimens had been dusted with arsenic.

29 May 2013

Around lunch time I turned the car towards the west, as I was heading to San Francisco via the tall-grass prairie of Kansas, the Rockies in Colorado and California's Yosemite National Park. It was 820 miles back to the Mississippi and I crossed it at St Louis, just below where the Missouri and Mississippi merge, at around the same time on 30 May. The easy driving, along long flat roads, turnpikes and freeways, gave me plenty of time to think about Passenger Pigeons and what I had learned over the last two weeks.

I had refreshed my appreciation of the American people. Everywhere I went I had been made to feel welcome, and I was constantly reminded that visitors to this great country are welcome and appreciated. I had been assumed to be Australian more times than is comfortable in the spring of an Ashes series but that was easily compensated for by the number of times I had been told that I had a lovely accent – it can go to your head, you know. And all those 'Have a nice day's seemed much preferable to the sullen grunts that would be more likely to be their UK equivalents. America is welcoming to strangers.

My encounters had not been with the average American – by dint of my journey I had met more than my share of motel owners, waitresses, naturalists and gas station attendants. What a heady mixture. But apart from the professional biologists I met, pretty much none of these people knew what I meant by a Passenger Pigeon. There had been a few false-positives, people who were confused over carrier pigeons, but otherwise there was just Susan, the chatty waitress in Hardinsburg who had had a Passenger Pigeon in her garden, and then the historian Michael Federspiel in Petoskey, who really did know what I was talking about.

Fair enough, it's the usual issue of shifting baselines – every generation grows up accepting the world around them as 'normal'. We don't miss what we've never had, and there is no-one who has had a wild Passenger Pigeon in their lives for well over a century. This was what Aldo Leopold foretold, writing in the late 1940s: 'Men still live who, in their youth, remember pigeons. Trees still live who, in their youth, were shaken by a living wind. But a decade hence only the oldest oaks will remember, and at long last only the hills will know.'

The men and women who knew Passenger Pigeons are indeed gone, and they have not passed the remembrance down to their children and grandchildren because there are so many more important things to talk about than dead birds. But I am glad that I met some trees in Dysart Woods and the Johnson Woods State Preserve that will have remembered those flocks of birds snaking across the sky. Even though the trees could not share their memories, I felt a lot closer to those days having seen some small remnants of virgin forest looking much as they must have done in the heyday of the Passenger Pigeon.

Each day I had been in the USA so far, and for several weeks to come on my journey west, I saw Mourning Doves in pairs or small groups. They would fly past or fly up from the roadside, and I would see them many times a day. Each time I saw a Mourning Dove I thought 'Passenger Pigeon', and I tried to imagine the times when the Passenger Pigeon was the commonest bird in the world. When the Europeans first invaded, about two out of every five North American birds were Passenger Pigeons. But the species never made it onto the list of birds for my trip.

My Passenger Pigeon road trip, from Mississippi to Mississippi, was approximately 4000 miles. If the five billion Passenger Pigeons who used to occupy North America had lined up along the road to wish me well on my journey there would have been 1,250,000 of them per mile, or more than 700 per yard of the journey, but they were all gone. Completely gone. Gone and largely forgotten. And that leaves us with two questions. How were they made extinct? That's the subject of the next chapter. Would we want them back? We will address that in Chapter 7.

CHAPTER FIVE

No ordinary destruction

The Passenger Pigeon needs no protection. Wonderfully prolific, having the vast forests of the North as its breeding grounds, traveling hundreds of miles in search of food, it is here today and elsewhere tomorrow, and no ordinary destruction can lessen them, or be missed from the myriads that are yearly produced.

Ohio State Senate, select committee, 1857

When Martha died in her cage in Cincinnati Zoo between midday and 1 pm on 1 September 1914 the Passenger Pigeon moved from extant to extinct. It was an unusually precise and a very personal extinction.

It's usually difficult to pin down the exact moment of extinction of a species. Many species just fade away and eventually there is growing realisation that they might, finally, have given up the ghost and be gone. There are usually moments of hope, sometimes uncertain sightings, but eventually we face the fact that another species is lost from our planet. The Passenger Pigeon went through those stages too, but in the last few decades of the nineteenth century it was clear that the Passenger Pigeon was much rarer than before, and by the last year of that century the last wild sightings were dwindling away. Rewards were offered for information about living Passenger Pigeons, but no-one was paid, and by 1910 it was clear to all that the only Passenger Pigeons left on Earth, though billions had been present a century earlier, were those in captivity in Cincinnati Zoo.

For just over four years it was clear that a female pigeon, known as Martha, was the last Passenger Pigeon on Earth. People came to look at her in her cage, knowing that they were looking at the last of her line, and knowing that when she died there would no longer be a single Passenger Pigeon left. Never before, or since, was any extinction such a visible process.

Martha was the last of the billions. This chapter addresses the question of how the billions were reduced to hundreds of millions, to tens of millions, to millions, to hundreds of thousands, to tens of thousands, to thousands, to hundreds, to tens, and to Martha.

Let us begin by getting a clearer view of the trajectory of the decline towards extinction.

The accounts of Wilson, Audubon and King, of single flocks numbering between one and three billion birds, were all made in the first half of the nineteenth century. This doesn't mean that the species had not already declined by that time, but at least it shows that it was not remotely on its knees with its head bent and waiting for the executioner's axe.

By the early nineteenth century it seems likely that the Passenger Pigeon had indeed declined considerably, however, because there is a dearth of records of nestings from the most populated states in New England and in the Atlantic provinces of Canada, even though these were areas of pigeon abundance in the early days of the European invasion of North America. And the notion that the birds were not as common as they had once been is a recurring motif in writings about Passenger Pigeons from the mid-eighteenth century. As early as 1754, Peter Kalm noted the link between the spread of the human population and declining numbers of pigeons:

> In forests where there are human settlements or where the country is inhabited, only a few are to be seen; and as the land is being gradually cultivated by man, the Pigeons move further away into the wilderness. It is maintained that the cause of this is, partly, that their nests and young are disturbed by boys, partly their own sense of a lack of safety, and finally that during a great part of the year their food is shared by the swine.

In 1838 Caleb Atwater, historian and politician, wrote of Ohio that:

> Formerly the pigeons tarried here all summer, building their nests, and rearing their young, but the country is too well settled for them

now; and so, like the trapper for beaver, and the hunter, they are off into the distant forests, where their food is abundant, and where there is none to disturb them in their lawful pursuits.

This was the type of observation that cropped up with greater frequency through the nineteenth century, and in highlighting food availability and disturbance in his short remark Atwater has also pointed to the cause of the decline.

McGee described the decline as follows for Iowa:

From the early sixties the pigeon migrations declined. In the early seventies flocks of diminishing numbers continued to fly in spring ... then about 1876 these ceased and the Passenger Pigeon became extinct in eastern Iowa.

The *Pittsburgh Dispatch* reported on 26 March 1870, under a headline 'The slaughter of the pigeons in Pennsylvania', that:

Accounts from rural districts ... prove this to be what in olden times they called a pigeon year. In other words, the woods and fields, in this and two or three adjoining counties, are filled with millions of Passenger Pigeons. Twenty years ago it was no uncommon thing to see these birds passing over, so thick, as, like the Persian arrows at Thermopylae, to darken the sun; while thirty years ago, those great pigeon roosts, covering miles of forest, with the birds breaking down the limbs of trees with their crowded weight, were well known to our farmers. But as our woods have been cut down and thinned out, and the country settled up, the pigeons have been disappearing westward, though we still hear of their great flocks and roosts in Kentucky and Indiana. This year, however, they seem to have returned to us, in greater force, than at any time within the last twenty years.

Again, there is a suggestion here as to the cause as well as the timing of the decline.

By the turn of the century the Passenger Pigeon had gone and people were wondering at its loss, and about its cause. The *New York Times* carried an article in its Sunday magazine which described the Passenger Pigeon as:

> a bird that is to-day extinct, so far as anyone has been able to discover, although less than fifteen years ago it was abundant on this continent and to the people of this State was as familiar as sparrows are now. Its disappearance came as suddenly, one might say, as the snuffing out of a candle. One day in 1889 these birds were apparently as numerous as they had ever been within the memory of man. The next day they had disappeared, and no one has seen or heard positively anything of them since.

And also:

> One of the greatest of these [nestings] was in Sullivan County, this State [New York], and the beech woods of the adjacent Counties of Wayne and Pike in Pennsylvania not more than 100 miles from New York City. The last appearance of any wild pigeons as far east as that, however, was in 1876, when they occupied the beech woods by the million. They never nested east of the Allegheny River after that, but in the hemlock and beech forests of Cattaraugus County, in this State, and the Pennsylvania counties south of it, they roosted and nested as late as 1886.

In 1905, Edward Howe Forbush, writing in *The Auk*, described the Passenger Pigeon as being 'practically extinct in New England for twenty years', and in his *Birds of Massachusetts and other New England States* (1927) he characterised the former status of the Passenger Pigeon in New England as a 'Formerly abundant migrant and common to abundant local summer resident in all New England States.'

The records from Canada are interesting, as population declines are sometimes more obvious at the edge of a species' range than at its centre. George E. Atkinson described the situation in Manitoba in William Butts Mershon's 1907 book on the Passenger Pigeon. He quotes a variety of correspondents thus:

George A. Garrioch wrote: 'I was born in Manitoba and came to Portage la Prairie about 1853. I was then only about six years old, and as far back as I can remember pigeons were very numerous. They passed over every spring, usually during the mornings, in very large flocks, following each other in rapid succession ... The birds, to my recollection, were most numerous in the fifties, and the decline was noticed in the later sixties and continued until the early eighties, when they disappeared. I have observed none since.'

Angus Sutherland of Winnipeg: 'I was born in the present city of Winnipeg and have lived here over fifty years. The wild pigeons here were very numerous in my boyhood. They frequented the mixed woods about the city, and while undoubtedly many birds bred here, I remember no extensive breeding colonies in the province, and believe the great majority passed farther north to breed. About 1870 the decrease in their numbers was most pronouncedly manifest, this decline continuing until the early eighties, when they had apparently all disappeared, and I have seen only occasional birds since, and none of late years.'

W. J. McLean, formerly of the Hudson's Bay Company and then Winnipeg resident: 'I came to the Red River Settlement in 1860 and found the pigeons very plentiful on my arrival. The birds came in many thousands, and great numbers of them bred in the northeastern portion of the province through the district north of the Lake of the Woods and Rainy Lake, where the cranberry and blueberry are abundant. These fruits constitute their chief food supply, as they remain on the bushes and retain much of their food properties until well on into the summer following their growth. They also feed largely on acorns wherever they abound. The decline began about the early seventies.'

E. H. G. G. Hay, formerly police magistrate of Portage la Prairie: 'I came to the country in June, 1861, and found that the pigeons were abundant previous to my arrival ... They seemed most numerous in the sixties and began to show signs of decreasing about 1869 or 1870, and by 1875 they had all disappeared and I have only seen an occasional bird since.'

Charles A. Boultbee of Macgregor wrote: 'I have resided in Manitoba since 1872, and have taken pigeons as far north as Fort Pelly in the fall of 1874, but know nothing of them previously. In our district they usually made their appearance in the fall and fed upon the grain. They continued fairly numerous until about 1882, at which time we had to drive them from the grain stocks, but they then disappeared and only stragglers have been noted since.'

The last big nestings of Passenger Pigeon were reported from Wisconsin, such as that at Wisconsin Dells, in 1871 and 1882; Michigan, such as that at Petoskey in 1878; and the Allegheny Mountains in Pennsylvania through the 1870s and early 1880s. Schorger writes that 'What was left of the former hosts nested in Wisconsin in 1885, in Pennsylvania in 1886 and again in Wisconsin 1887.'

The picture emerges of a retreat of the Passenger Pigeon from those parts of its range where human population was highest and growing. The birds retreated from the onward march of what we call civilisation. The retreat started in the coastal states and provinces as they were colonised, and continued as the human population expanded and then spread westwards. The last bastions of the Passenger Pigeon were the more remote areas – those towards the western and northern edges of its range, and those wooded mountainous areas where human population densities were lowest.

The decline of the Passenger Pigeon probably occurred over the course of three centuries, and it most likely consisted of a huge contraction of its breeding range and then a final decline in population despite the apparent continuing availability of suitable nesting sites. That is the phenomenon that we need to understand and explain.

THE FOUR HORSEMEN OF THE ECOLOGICAL APOCALYPSE

Jared Diamond, in a review of extinctions he published in 1989, referred to the four ecological Horsemen of the Apocalypse that have ridden through the land of species extinctions. They are not Conquest, War, Famine and Death but Invasives, Habitat Destruction, Overkill and Secondary Extinctions. Which one of these dealt the blow to the Passenger Pigeon? Whichever horseman was responsible he must be a fearful and powerful foe to reduce the commonest bird on the planet from billions of birds to a sad lonely female in Cincinnati Zoo in the space of a century.

In assessing the roles of Diamond's four horsemen we should keep five things in mind. First, the biology and demography of the Passenger Pigeon, as outlined in Chapter 2. This was a forest species that relied on tree mast for successful nesting, and successful nesting more than once a year was needed to maintain the population.

Second, since we are dealing with the whole world population of a species, unlike when dealing with a local population, we can ignore the effects of immigration and emigration, although it is perhaps worth noting, just in passing, that one proposed explanation for the disappearance of the Passenger Pigeon was that they had all flown off somewhere – the Moon, the other side of the Rockies and South America were all suggested.

Third, in discussions of the Passenger Pigeon's extinction there has been a tendency to concentrate on the impacts on the breeding grounds, and most particularly in those last breeding grounds in the last years of the species' existence, and to neglect the less dramatic and less obvious factors and any that pertain primarily to the wintering grounds. We should be prepared to consider factors that act at any time of the year and in any part of the geographic range of the species.

Fourth, the factors leading to extinction may have varied in nature or strength through the course of the species' decline. There is no reason why the cause of the decline in the first half of the nineteenth century should necessarily be the same as that which finished the bird off between 1850 and 1900 (it may have been – but we should keep an open mind).

Fifth, it is almost certainly wrong to look for 'the' cause of the bird's extinction – anything that detrimentally affected its vital rates of survival and reproduction may have played a part in the decline. We may be looking for a suite of answers rather than just one.

Horseman 1: Introduced species
Introduced species (including really small species that we call diseases) are a very potent threat to the world's biodiversity. Taking just birds, Birdlife International assesses 820 of the world's *c*.10,000 bird species as being threatened by invasive species and diseases. This aspect of life and death rarely crops up in discussions of the reasons for the extinction of the Passenger Pigeon, so perhaps we should just skip it here – but on the other hand, maybe we shouldn't.

Although introduced predators – from rats on seabird islands throughout the world, and cats attacking flightless species in New Zealand, to tree snakes in Guam – have been potent causes of avian extinction, there is no suggestion that a new predator was introduced onto the American continent that affected Passenger Pigeons.

Schorger devotes fewer than two pages to the subject of disease and is clearly not that enamoured of the suggestion. We now understand, however, that diseases and parasites have been powerful forces in shaping bird behaviour, plumage and ecology – and there can be few species whose ecology and behaviour would more certainly favour the spread of disease than the Passenger Pigeon. It fed, migrated, nested and slept in huge flocks where the birds were pressed close together. In winter roosts the numbers were so enormous that the birds roosted on top of one another, and in the breeding season individual trees often held scores of nests, with all the potential for droppings from above to come in contact with nests and birds below.

It may be that the Passenger Pigeon had a cast-iron constitution and was, perhaps literally, immune to all challenges from diseases, but that is as conjectural as imagining that disease might have played a role of in its demise. E. T. Martin, a game dealer of Chicago, who might not be considered to be entirely neutral on the subject given that his profession had a role in the demise of the species, was keen on the idea that disease was an important factor. He wrote that 'pigeons in captivity were very susceptible to disease' (through his profession he was in a good position to know this), and of a case where:

> over 20,000 of them were penned in rooms sixteen feet square – a thousand to each room. They had cleaned the mud off their feathers, were eating well and appeared as strong, healthy a lot of birds as one could wish to buy. One morning they were fed as usual at sunrise, and the birds in every pen ate their half bushel of corn, then looked for more. An hour later nearly all the pigeons in number one room were dead or dying of canker. Before a man could return from downtown with sulphur and alum – a trip of an hour – the birds in the next pen were dying rapidly, and some were dropping from the perches in room number three. Prompt action checked the disease there. Had no remedies been within reach nearly every one of the 20,000 pigeons would have died inside of a few hours.

Schorger questioned the diagnosis of canker (trichomonosis), but the case is interesting because of the rapidity with which birds were dying (whatever they were dying of) and their rapid recovery when treated (which they wouldn't have been in the wild). Canker is a disease commonly found in doves and pigeons (but responsible for a recent population decline in Greenfinches in the UK). It is caused by a protozoan, and is more easily transmitted to their young, perhaps through the ingestion of pigeon milk in the first days after hatching.

Martin queries, and Schorger stresses, the fact that there do not appear to be records of people noticing die-offs of pigeons at colonies, and this is a fair point. However, many colonies were not accessible, and the people entering colonies were mostly in search of live pigeons to kill, so they may not have remarked upon even thousands of dead birds. Moreover, there are many fewer reports of the winter roosts, and disease may have hit the birds harder in winter than in the breeding colonies. William W. Thompson, who was quite an adherent of the possibility that disease was a factor, wrote that he always saw many dead birds when he entered nesting colonies in Pennsylvania.

The growing ranks of European invaders through the nineteenth century brought with them domestic pigeons, which may have brought in diseases new to the American continent in the same way that the humans brought smallpox, measles, scarlet fever, diphtheria, typhoid and a host of other diseases to which the Native American peoples had not been previously exposed (and which killed millions of them). Domesticated pigeons had been present in the USA for three centuries or more without, as far as we know, causing large losses to Passenger Pigeon populations, but the new arrivals might have brought new diseases, and old diseases can sometimes evolve into more lethal strains (as we have seen recently in the case of bird flu and other strains of influenza), and this may have happened at a crucial time for the Passenger Pigeon – although this is merely conjecture.

Also, at this time Europeans were introducing European bird species to the Americas. House Sparrows were introduced to the USA in the 1850s and they spread rapidly across the continent. This is just the type of species which would interact with Passenger Pigeons in fields and around farm buildings, and indeed the two may have competed for food supplies to some extent. House Sparrows don't eat acorns or beech mast but the Passenger Pigeon took a whole range of seeds of agriculture, and those are also main elements of the diet of rural House Sparrows. There is no shred of evidence known to me that suggests that the spread of the House Sparrow led to transmission of diseases to native birds, but disease transmission is a very difficult thing to pin down and it would be rash to rule it out completely. The arrival and spread of the House Sparrow coincides very well with the accelerating demise of the Passenger Pigeon.

There are early records of the pigs of European settlers, which fed largely on tree mast, starving or finding existence difficult if their areas were visited by large flocks of feeding pigeons. But this interaction must have worked both ways. The Passenger Pigeon's dependence on beech mast, acorns and chestnuts would have made it vulnerable to the great rise in the population of pigs, which would have been scouring the same woods for the same rich nuggets of protein, carbohydrate and fat.

Ohio, Illinois, Indiana and Kentucky were the main areas for this industry, which reached its height in the first half of the nineteenth century. From 1835, Cincinnati was nicknamed 'Porkopolis' because of its role as the pork-packing centre for the country. Pigs were turned out into the woods for two years at a time and then rounded up and fattened on corn. André Michaux, the French botanist and explorer, wrote of Kentucky in 1802: 'Of all domestic animals, hogs are the most numerous; they are kept by all the inhabitants, several of them feed a hundred and fifty or two hundred. These animals never leave the woods, where they always find a sufficiency of food, especially in autumn and winter.' Apparently razorbacks were as common in the forests of the Midwest as 'grains of sand on the sea-shore'.

Although there is little direct evidence that supports disease as being important in the decline towards extinction of the Passenger Pigeon it would be foolish to write it off completely, and the impact of large numbers of free-ranging pigs in the breeding and wintering range of the birds must have reduced the food available to them to an appreciable extent. We shouldn't think that canker and hogs were the reason for the Passenger Pigeon's extinction – but nor should we rule them out of having had any impact at all.

Horseman 2: Chains of extinction
We talk about food chains and food webs, and those concepts point to the fact that species are linked in many ways. Imagine that there were a species that lived on nothing but Passenger Pigeons; when the Passenger Pigeon ceased to exist then so would that species. If one species suffers losses of population then others with which it interacts might either suffer or benefit, and these interactions could lead to those or other species in the chain being affected.

It was formerly thought that the Passenger Pigeon's extinction took another two species down with it – two species of feather lice that were thought to live literally *on* the Passenger Pigeon. One of these, *Columbicola extinctus*, was brought back from extinction by Dale Clayton and Roger Price when they published a paper in 1999 showing that it was the same species as a feather louse living on Band-tailed Pigeons in the USA (and the world sighed in relief, knowing that the louse was safe). The other, *Campanulotes defectus*, is now considered to be the still-extant *Campanulotes flavus* of Australia, and so this was probably a case of within-museum contamination of a Passenger Pigeon specimen.

For all we know, some particularly interesting species of parasitic intestinal worm may have perished with the Passenger Pigeon, but if so its loss goes unrecorded and unmourned.

What seems unlikely is that the Passenger Pigeon's own extinction was part of some domino effect of knock-on species extinctions. However, on my journey around the USA I did experience a fleeting increase in heart beat when I wondered whether the fate of the American Chestnut and the Passenger Pigeon were in some way interlinked.

Passenger Pigeons ate chestnuts, and Schorger regarded the American Chestnut as the bird's third most important tree food after beech mast and acorns. The American Chestnut was a widespread and common species through much of the breeding and wintering range of the Passenger Pigeon. But now there are only a few hundred mature specimens, and although saplings are widespread they all die at an early age. What happened to the American Chestnut is very similar to what happened to the English Elm in England – an introduced fungus wiped it out over a matter of a few decades. Did this play a role in the Passenger Pigeon's demise? We can be sure it didn't, as the disease, chestnut blight, arrived in 1904 – by which time Passenger Pigeons existed only in captivity in Cincinnati Zoo.

Horseman 3: Overexploitation

Passenger Pigeons were good to eat, sometimes ridiculously easy to capture, fun to shoot and had feathers that some believed would give a long life to those who slept on pillows stuffed with them.

Everyone wanted to get their hands on them – and because their appearances were unpredictable in any location it must always have felt like an opportunity that should not be missed. The overwhelming abundance of the bird when it arrived in a locality made it seem as if an ocean of birds were present – and no-one can empty the ocean.

Before the arrival of the European invaders, and for thousands of years, the Native Americans had harvested Passenger Pigeons at their colonies as well as at their roosts. Each year scouts would be sent out to see if there was a pigeon colony within easy reach, and if so then whole villages would move to feed off the fat of the land. Native Americans, at least some of them, had strict rules about only taking the squabs from the nests and leaving the adults alone, which led to disagreements with the Europeans in later years when such conservation measures were ignored. We perhaps should not give the Native Americans too many marks for ecological know-how and prudence as they were limited in the number of pigeons they could take by their weaponry of sticks and bows and arrows, but their approach to harvesting squabs looks pretty wise compared with the profligacy of the approach taken by the Europeans.

By limiting the killing of adults, and concentrating on the young birds, Native Americans reduced the chances that their activities would cause the adults to desert the colony – the last thing that they would have wanted, as the colony would provide food for many people for weeks at an end. Also, reports suggest that the meat of the adults, though perfectly edible and nutritious, could be tough and rather dry, whereas the fat young pigeons were said to be delicious.

Native Americans shot the squabs out of the trees with blunt-tipped arrows and used long poles to push them out of the nest. Native American villages would keep the fat from squabs as a kind of butter for use through the coming year, and thousands of pigeons would be dried for winter food.

Pigeons were an occasional bonanza for hard-up European invaders. Passenger Pigeons would be hunted and shot, trapped and even bashed out of the air with sticks and oars. This was not the sort of hunting that most people were used to. You didn't have to stalk Passenger Pigeons or know your quarry intimately – this was a species for the amateur hunter. All you had to do was notice that the air was full of pigeons and fire into the throng to get your lunch and dinner, and maybe enough for a few days at that.

There are accounts of 30, 40, 50, 80, 120, 124, 130, 132, 143 or even 'two to three hundred' individual pigeons being shot out of a passing flock with a single discharge of a gun. Even if these were exaggerated accounts, or at least at the very highest range of what was ever possible, they indicate the abundance of the birds in flocks, and something of the density of birds in those flocks.

Nineteenth-century USA was a country of armed men. The Bill of Rights in 1791 had introduced the Second Amendment to the Constitution, which guaranteed the right to keep and bear arms. No surprise, then, that when the pigeons arrived there were plenty who were armed and ready for them. And there were no limits, except your own ammunition supply and conscience, on what you could shoot or when you could shoot it. An adult heading north to nest in spring was no less fair game than a young bird heading south in its first winter. The Passenger Pigeon was just the type of species where it was easy to fool yourself that your killing couldn't be making much of a difference, and where the temptation to cash in on a temporary local glut of birds was overwhelming.

The Europeans also trapped Passenger Pigeons for food. The birds were caught wherever they could be attracted to the ground in large numbers, such as at naturally salt-rich localities, bare earth baited with salt, cereal grains, acorns or beech mast – and they were also caught by using live captive pigeons as decoys. Stool pigeons were captive birds, often kept for many years and tamed, which were tied to a stool that could be lowered and raised. When the stool was lowered a good stool pigeon (and not all captive Passenger Pigeons made good stool pigeons) would flutter, resembling a bird about to alight, and this would attract passing flocks of wild pigeons. 'Flutterers' would also be used – Passenger Pigeons allowed to fly but tethered on long strings. Both flutterers and stool pigeons were often blinded by having their eyelids sewn shut.

A variety of fixed, hand-held and spring-powered nets were used to catch pigeons attracted in these ways, and scores or hundreds of birds could be caught in one spring of the net. A single day's haul from just one net could amount to thousands of birds, and some netters caught tens of thousands during a season.

Some birds were simply killed (and plucked and preserved, or eaten fresh), but others were kept in captivity and fattened up, and still others were destined for another end. Because of their strong flight, Passenger Pigeons were used in shooting competitions which were the forerunners of the rather less bloody sport of clay-pigeon shooting. Large numbers of birds were used in these shoots, with the pigeons held in cages ('traps') from which they were released into the air as targets for competitors. To persuade the birds to fly, various ingenious means of propulsion were tried such as mounting the traps on bent sticks and releasing them and opening the door at the same time so that the bird was hurled into the air, but this often meant that the pigeon was a helpless target until it righted itself and gained control, and this was not considered good sport. Thousands of birds could be killed in a day.

Trap-shooting produced new heroes, most notably Captain Adam H. Bogardus, who won many trophies, bets and competitions. He once shot 500 pigeons in well under the 11 hours required to win a wager, and he was the first person to shoot 100 pigeons consecutively (without a miss) from traps. Trap-shooting began in the mid-1820s and lasted into the 1880s, with its peak in popularity coming in the 1860s and 1870s. This seems to us these days, as it did to a growing number of people in the mid-nineteenth century, to be an abomination. Passenger Pigeons from the 1871 colony at Wisconsin Dells were transported to trap-shooting contests in New York State. A matter of hundreds of miles, just to be living targets for marksmen to demonstrate their skill. Public revulsion grew over this practice, and the shooters turned to glass balls (which were more satisfying when they were filled with feathers) and eventually to the blue-grey clay pigeons that are used today.

But even though trap-shooting involved tens of thousands of live pigeon targets a year they were being taken from a population numbering hundreds of millions at the time. They should be no more than a footnote to the story of the Passenger Pigeons' demise.

Most of the Passenger Pigeons that were trapped or shot were destined not for sport but for immediate consumption. About half a dozen pigeons would fit in a pie, which would often carry three pigeon legs in its crust to show what delights were contained within.

Etta Wilson (1857–1936) was a journalist and writer, and a keen observer of birds, who grew up in the upper part of the lower Michigan peninsula where her father (a Native American) was an accomplished pigeon hunter and provided plenty for her mother, the daughter of missionaries, to prepare for the family. Wilson describes the domestic scene thus:

> Many Pigeons were eaten each day throughout the season. In our large family four dozen or more could be eaten in one day. The birds were not large and a husky boy could devour three at one meal. Always desiring to emulate my brothers in anything they might do, I ate three Pigeons at dinner one day but the after effects were so upsetting that they nearly robbed me of my Pigeon appetite for that summer.
>
> Ordinarily we stewed or baked the birds, adding quite a large piece of salt pork to the old birds since they were never very fat. Mother had a big pot and into it she would pack as many birds as it would hold, cover them with cold water, add the pork cut into small pieces and sliced potatoes, season with salt and black pepper and stew for about forty-five minutes. Or she would place a similar combination in a big dripping pan and bake in the huge elevated oven, which was capacious enough to cook a meal for our big family.
>
> Young birds were so enormously fat that we usually broiled or fried them but it was a long task to prepare enough birds in this way for so many persons. Less time was required to cook young birds, they were extremely tender, but I never ate a Pigeon of any age that was not delicate and delicious. The meat is darker than the dark

meat of a chicken and is entirely without strong taste. When stewed the meat separates readily from the small bones and every part of the cooked bird may be eaten.

At an immense colony in southern Canada in 1835, 'wagon loads of the young birds could be easily obtained', and no doubt they provided food for the local people for a goodly time. But a lot of the exploitation of the Passenger Pigeons at this time would have been for local use. In an era when refrigeration depended on how long your ice would last and transport was limited to the quantities and distances that could be accomplished by animal-drawn wagon, the numbers that could be taken for local use might have been large in absolute terms, but they were low in proportional terms when we consider that the colonies would be numbered in millions of birds. But the human population grew rapidly within the breeding and wintering range of the Passenger Pigeon in the nineteenth century, and therefore the impacts of even purely local use would have increased steadily and remorselessly. Although the pigeon's biology had evolved to swamp the impacts of local predators, humans were pretty determined predators, and their numbers were increasing rapidly.

And, as predators, the humans became far less local too. The spread of the railroad and also the telegraph allowed a deadly combination of events to unfold. There were no legal restrictions on hunting Passenger Pigeons or most other wildlife, and so nesting colonies were 'fair' game and provided relatively easy pickings, as long as you could find out where they were (which was aided by the telegraph) and get to them (which was aided by the railroad). The railroad also opened up vast areas of the country as a market to be supplied with cheap meat. This combination of human advances in population, communication and transport, coupled with the lack of protection for wildlife and the vast assemblages of Passenger Pigeons, allowed the supply of pigeon meat to become an industry rather than a rural pursuit.

Game dealers were the essential link in the chain. They negotiated deals to supply game markets in eastern cities, arranged transport for the live or plucked birds, and bought produce off the netters and shooters. This was an organised activity as far as the people involved in it were concerned – from wood to railroad station to game market to restaurant or kitchen – but a chaotic one from the Passenger Pigeons' point of view.

One of the earlier examples of the industrialised slaughter of Passenger Pigeons took place in 1871 at Kilbourn City (now Wisconsin Dells), where I passed through on 23 May 2013, and in 1878 another occurred at Petoskey, which I visited a day later.

Jennifer Price writes that the railroad arrived at Kilbourn City in 1857 and reached Black River Falls in 1870. In 1871, when the Passenger Pigeons arrived to nest in the largest single colony that we have had described to us, of around 150 million birds, with more spread over the general area, there was a railroad at either end of the colony and at Sparta in between. The pigeons made their own way to this area but millions of them were shipped out on the railroads that had also brought in hundreds of hunters. Barrel-loads of pigeons were loaded onto the trains and shipped east to Milwaukee, Chicago, St Louis, Cincinnati, Philadelphia, Boston and New York.

Schorger, based on newspaper accounts of the time, estimated that the trapping season lasted for around 40 days, during which time 100–200 barrels daily were being transported. Extra trains, running at night, were needed to cope with the cargoes to be transported. Each barrel contained up to 30 dozen pigeon bodies, suggesting a total shipment of pigeons of around two million. More were killed and left to rot or were wounded and died later. Thousands and thousands of squabs would have died in their nests as one, or perhaps even both, of their parents were killed. But this was from a colony which almost certainly numbered more than 100 million nesting birds.

In 1878 there were huge numbers of pigeons nesting near Petoskey, Michigan, where the railroad had arrived a few years earlier. Again, by boat and train more than one million Passenger Pigeons are thought to have been sent to the distant markets, and more would have been consumed locally – and again, many must have been injured and many squabs must have starved in the nests. This colony covered around a quarter of the area of the Kilbourn City colony of 1871, and therefore the potential impact of the hordes of pigeon hunters and shooters would have been higher than at Kilbourn.

Etta Wilson gives an account of the killing in this area:

Day and night the horrible business continued. Bird lime covered everything and lay deep on the ground. Pots burning sulphur vomited their lethal fumes here and there suffocating the birds. Gnomes in the forms of men wearing old, tattered clothing, heads covered with burlap and feet encased in old shoes or rubber boots went about with sticks and clubs knocking off the birds' nests while others were chopping down trees and breaking off the over-laden limbs to gather the squabs. Pigs turned into the roost to fatten on the fallen birds added their squeals to the general clamor when stepped on or kicked out of the way, while the high, cackling notes of the terrified Pigeons, a bit husky and hesitant as though short of breath, combined into a peculiar roar, unlike any other known sound, and which could be heard at least a mile away. Of the countless thousands of birds bruised, broken and fallen, a comparatively few could be salvaged yet wagon loads were being driven out in an almost unbroken procession, leaving the ground still covered with living, dying, dead and rotting birds. An inferno where the Pigeons had builded their Eden.

The Passenger Pigeons returned to Kilbourn City in 1882, and this time the slaughter appears to have been unremitting in its wastefulness. In mid-May more than 15,000 squabs had to be buried because the dealers had bought more than they could preserve, and several tons of dead birds were ditched in the Wisconsin River for the same reason. The stench of rotten squabs must have been appalling.

The plundering of Passenger Pigeons in Michigan and Wisconsin in the last few big nestings was the antithesis of sustainable harvesting. Breeding colonies were disturbed and adults and young were taken with no control on numbers and in the most wasteful way. As David Blockstein and Harrison Torduff, writing in *American Birds* in 1985, pointed out, it wasn't just the number killed that was important in these later years, but the huge disruption to the nesting attempt, which would have drastically reduced productivity. Very few young can have fledged from some of the last big colonies, and for several years Passenger Pigeons not only failed to have two successful nesting attempts, they didn't even have one.

Mershon wrote that 'the old birds were netted continually, and had no chance to rear their young,' and that 'the birds that were left after the general extinction of the mighty host failed to reproduce their species and soon became extinct.' The breeding failures brought about by disturbance from hunting pressure do seem to account for a large part of the role of overexploitation in the story of the Passenger Pigeon's decline in numbers.

Generally speaking, we are keen to blame the man standing at a colony and aiming his gun (although not much aiming was needed) at a flock of pigeons, because it is clear that he caused deaths directly. But the unsustainable carnage of plundering the colonies was just one part of the final chapter of the Passenger Pigeon story, not its main theme. The early harvesting of colonies by Native Americans was sustainable, we must assume, because the population density of the Native Americans was low and the take of birds was to some extent regulated by tradition and custom. The main take was of squabs, and the adults were left alone in order to ensure that a good harvest of squabs could be taken.

The exploitation of the Passenger Pigeon in its last century became more important as the human population taking part in the slaughter increased. This probably applied to those targeting the winter roosts and the flocks that passed over the country, as well as to the growing band of itinerant 'pigeoners' with their nets and guns who waited for news of a new nesting and then travelled *en masse* to exploit the riches. The direct pigeon death toll grew as the human population grew, but also as the effective 'local' population grew through improved transport links. The number of mouths within two days of a Passenger Pigeon colony increased from thousands to millions with the nineteenth-century expansion of the railroad. Passenger Pigeons killed in Wisconsin were eaten in New York and Boston in the 1870s and 1880s – this could not have happened in the 1820s or 1830s.

Horseman 4: Habitat loss
The Passenger Pigeon was a bird of the eastern forests. It nested in trees, it roosted in trees, and its whole lifestyle was shaped by the variations in abundance of the tree mast that formed a large part of its food throughout the year, as well as the food on which it fed its young. What of the trees? What were the forests of eastern North America like before the European invasion?

The main trees of the breeding range of the Passenger Pigeon were American Beech, a variety of oaks and American Chestnut. Both beech and oak have occasional years of producing bumper crops of mast or acorns, and it was these years of plenty that shaped where the Passenger Pigeons nested. Only large areas of forest with abundant supplies of tree seeds could support colonies of tens of millions of Passenger Pigeons, scouring the woodland floor in great flocks for tree seeds in their various shapes and forms.

Beech trees only begin to produce mast when they are 40 years old, and not until the tree is 60 years of age does mast production peak. The Red Oak starts its acorn production at around 25 years of age, but not abundantly before 50 years of age (White Oak peaks at 50 years; Bur Oak at 35 years). Clearly, mature trees were important to Passenger Pigeons, as they were the ones that produced abundant seeds – but not every year.

An individual beech tree produces high quantities of mast once every 2–8 years, and so do all the other beech trees around it in those same years. Masting is influenced by the preceding couple of years' weather, which will be synchronised over a large area. Red Oaks have acorn years every 2–5 years, and acorn production is also synchronised over large areas in that species. The least variable of the key tree species is the chestnut, which produces seed in large quantities every year or every other year.

Seed production by these trees varied from very low in non-mast years to very high in mast years, with beech producing up to 2,800 kg/ha, Red Oak up to 700 kg/ha, and American Chestnut up to 400. As far as calorific value is concerned, American Beech (550 Cal/100 g) outranks Red Oak (490 Cal/100 g), which outranks American Chestnut (375 Cal/100 g).

Taken together, these two sets of figures make it easy to see why the three main trees were ranked in the order that they were: areas of beech trees which had masted were worth seeking out because the seed supply would far outstrip those of other trees and the seeds were richer in energy. American Chestnut was less calorifically rewarding but much more predictable, and oaks were in between.

Oaks were the dominant trees in Wisconsin, Illinois, Missouri and Pennsylvania, whereas beech was dominant in Indiana, Michigan, New York, Vermont, New Hampshire and Maine.

One account mentions that nesting of Passenger Pigeons only occurred in Michigan in the years 1868, 1870, 1872, 1874, 1876 and 1878, and a similar pattern was noted in Potter County, Pennsylvania, in 1868, 1870, 1876, 1878, 1880, 1882, 1884 and 1886. C. Hart Merriam noted mast years in the Adirondack region of New York in the odd-numbered years of 1871, 1873, 1875, 1877, 1879, 1881 and 1883. Whilst this degree of regularity seems too good to be true, there is certainly a suggestion that nestings would take place about every other year and after years of high mast production.

Donald E. Beck, of the US Department of Agriculture Forest Service, measured acorn production in the southern Appalachians of North Carolina (within the former wintering range of the Passenger Pigeon) in the 1960s and 1970s and showed that White Oak had mast years in alternate years but northern Red Oak, Black Oak, Chestnut Oak and Scarlet Oak were much less predictable. Black Bears in New York have cubs in alternate years to coincide with mast years of oaks.

There are few remaining areas of original forest surviving in the range of the Passenger Pigeon. Despite the fact that many of the states where it wintered and nested are well-clothed with forests, these are secondary forests – ones which have regrown after the original trees were felled. Forests were felled for timber for houses, for firewood, for furniture and for railroad sleepers, and also to free land for growing crops.

Stuart Pimm and Robert Askins estimated 1872 as being the time when forest cover in eastern North America reached its lowest ebb – at around 48% of the original, pre-European-invasion level. Let us imagine what that meant for the Passenger Pigeon, and first let us consider the wintering population, as that often gets scant mention in this story.

The population levels of many temperate-latitude bird species are set during the winter – winter is a tough time with harsh weather conditions which increase metabolic rates, and day lengths are shorter so the time available for foraging is reduced. On the other hand, birds are not tied to nesting locations and so can roam widely looking for food, and we know that Passenger Pigeons were efficient foragers in this way.

If half of the trees on which you depend for food are cut down then your food supply has halved – and that was the position in which the Passenger Pigeon found itself on its wintering range in the early 1870s. And it wasn't just that every other tree in the forest had disappeared, the forest losses were clumped in space. The lowland areas nearest to the eastern seaboard, with highest human population levels, suffered most deforestation, and the more western and more mountainous areas were spared until later in the period. The Great Smoky Mountains, created as a National Park in 1934, was only selectively logged up until the 1880s, but then the deforestation of other areas made logging of this remote mountainous forest economically worthwhile, and by the late 1920s about two-thirds of the area had been logged.

For the Passenger Pigeon the geographic pattern of forest loss would have been very important. We can imagine that the whole of the wintering range of the species might have been covered by an individual bird during its lifetime. One winter the forests of South Carolina and Arkansas might have had good feeding conditions, and in other winters perhaps it was Tennessee and Louisiana. In the years when the oaks and beech of Arkansas aren't producing beech mast and acorns you need there to be forest in Alabama where you can feed. The Passenger Pigeon needed to range over huge areas to find enough food, and it managed to cope by seeking out the places where mast had been produced in adequate quantities. With large-scale, spatially contiguous deforestation then the number of options for feeding would have been reduced – it's no good if there are lots of trees but they all happen to be in places of low mast production, and that would increasingly have become a factor, from year to year, as deforestation proceeded.

Allied to this, the figure of 48% deforestation is an overall figure for forest cover. We can be sure that even in those areas where there was forest cover, many of the older trees would have been selectively logged – and these would have been the most productive in terms of mast. A forest of young trees was no use to Passenger Pigeons looking for food. Thus we can assume that a halving of forest cover on the wintering grounds would at least have led to a halving of the pigeon population, but it may in fact have had a far greater impact. On its own, and without any other factors, it could have reduced the population by as much as 75–80%, although that is a range plucked out of the air by my ecological intuition rather than a well-worked figure (but after all, that is what 'experts' are for).

If we turn our attention to the breeding season now, the same factors will come into play – except that the Passenger Pigeon would have been even more vulnerable because now it was tied to a particular location for nesting. You can't roam that far if you have to incubate an egg or keep returning to a colony to feed your single youngster. Large areas of the former range, whilst still having some tree cover, were rendered useless because there simply wouldn't have been enough trees within reach of a large colony to meet the demands of so many birds.

It was no accident that the last major nestings of the Passenger Pigeon were in the remote Midwestern parts of the range in Wisconsin and Michigan, where appreciable areas of virgin forests persisted after 1850 (though they were being lost increasingly rapidly as time passed) and in the most mountainous areas of Pennsylvania. In 1840 the centre of the lumber industry was New York and Pennsylvania, but by 1880 this had moved to Michigan. From about 1840 to 1900, most of the Michigan forests were cut down for farms and to produce lumber for buildings, ships and mines.

States such as Ohio would have been very largely covered with forests at the time of the European invasion (around 95%); it was only 54% forest in 1853, 18% in 1883, and by 1900 only about 10% of the state was forested. The decline of virgin old-growth forest would have been even more rapid, and mature trees would have been removed from those forests that remained. The remaining forest would have existed in small patches where Passenger Pigeons attempting to nest would have been more vulnerable to natural predators and local exploitation by the growing human population. There was nowhere in Ohio by the middle of the nineteenth century where Passenger Pigeons could nest in unbroken colonies 10 miles in length and several miles wide: there were no such forests. The quote from Atwater (in 1838) earlier in this chapter suggested that Passenger Pigeons ceased to nest in Ohio by that time, and this should not surprise us. Even had no-one ever shot and eaten a Passenger Pigeon, then I believe they would have been lost from Ohio by the 1850s because of habitat destruction.

And what happened in Ohio had happened in the more easterly former range states even earlier. George Peterken, in *Natural Woodland*, describes forest cover in Massachusetts as a whole and in the area of the town of Petersham, and within that area the Prospect Hill tract. These days the Prospect Hill tract is around 100% wooded, as it was in the 1720s, but the low point of forest cover was reached in 1860 when only 15% of the tract was a forest. Given that beech and oak do not produce many seeds until over the ages of 40 and 25 respectively, the trees of Prospect Hill, despite reaching a coverage of around 70% by 1900, were of little or no use to Passenger Pigeons for food production until about the time of the species' extinction in the wild. Very similar figures apply to the local area of Petersham, and to Massachusetts as a whole, in which the low point of forest cover, about 40%, was reached in about 1870. Gordon Whitney describes similar dramatic declines in tree cover in different areas of Massachusetts, in New York, and in Rhode Island. Much of New England would have been the same – few trees, even fewer old trees, and lots of people with guns. These states were no places for pigeons, and the march of civilisation pushed the birds further and further west and north into states such as Ohio, which in their turn became unsuitable – and the pigeons were left with few options for nesting.

Under these circumstances the infrequent mast production of the trees must have become more of a factor. When practically all of eastern North America was covered by forest the Passenger Pigeons could always find areas which were rich in mast. If the mast years in, say, New York, were in odd-numbered years then the pigeons, in their nomadic way, needed to find mast elsewhere in even-numbered years. Before deforestation, Ohio or Massachusetts might sometimes have provided abundant food, but with the forests gone the birds had to look elsewhere, and sometimes they might have failed. As the areas of suitably large areas of forests declined, then increasingly there must have been some years when the remaining forests all had non-mast years, which would have led to non-breeding or massive breeding failure.

Because the Passenger Pigeon needed to approach two successful nesting attempts each year to hold its own in population terms, then the impact of deforestation becomes even clearer. Was there actually enough forest in large chunks that were rich in tree mast for Passenger Pigeons to nest successfully, twice, in large numbers every year? The answer appears to be 'no'. Deforestation made most of the former range states unsuitable for large-scale nesting.

The aggregated impacts of loss of forest, loss of old trees and fragmentation of forest cover, throughout the bird's range in summer and winter, must have been huge. This was a forest-dwelling bird which had ranged freely over continuous tracts of forest for thousands of years – seeking out the hotspots of seed production. Those forests were rapidly reduced in area, and through the nineteenth century the flocks flew mostly over fields seeking out the remnant forests and checking them for beech mast and acorns. There was less and less chance of fitting in two nesting attempts in the season, and in some years it was probably the case that only a small proportion of the pigeon population nested at all.

THE REMAINING RIDDLE: THE LAST FEW

The last knotty bit of the extinction conundrum is the fate of the last few Passenger Pigeons. After around 1885 there was little commercial hunting, because the birds were too thin on the ground; there were laws passed protecting the Passenger Pigeon, which might have put off some hunters; and the population appeared, though it is difficult to say, to be so low that surely it was no longer limited by the quantity of tree nuts available. Why did the last 100,000, the last 10,000 or even the last 1,000 Passenger Pigeons die? Why did they not stage a recovery, and why are they not still breeding in much reduced numbers in parts of eastern North America?

Many species of birds have been driven to the very point of extinction and yet have bounced back from tiny populations to make a recovery. The Chatham Island Robin was reduced to a world population of five individuals, including just one fertile female (known as Old Blue), but now numbers more than 200 birds thanks to intervention by the New Zealand Wildlife Service (and Old Blue deserves some credit too).

Populations of most species, certainly most vertebrate species, tend to fluctuate within limits, and a process called density dependence has widely been shown to operate amongst birds and many other animals and plants. If a population increases one year (perhaps due to a good breeding season or lowered predation impacts) then the members of that population find life a bit tougher the next year – there is increased competition for food and/or for nest sites, and/or disease takes a bigger toll – and the population tends to decline again. Similarly, if the population declines for some reason (such as in a hard winter or after poor breeding success) then the members of that population find life a little easier the next year and the population tends to bounce back. The average level of the population is determined by the overall level of resources (usually food supply, but for some species nest sites if they are very specialised, and a little bit by predation effects).

This density dependence is a form of negative feedback that tends to regulate populations around a long-term average. It is not magical, it is just how things tend to work. Its operation means that it should become more and more difficult to kill off a species as its population becomes smaller and smaller. And yet, of course, species are made extinct. Most avian extinctions since the Passenger Pigeon have been species with restricted ranges, such as those species living on single islands, and they have been driven to extinction mainly by their whole habitat being chopped down or by overexploitation by humans or through the introduction of exotic predators against which they have no defences. Among these extinctions, the Passenger Pigeon is unusual in being a continental species with a large range (and a very large population). We need to identify some very powerful forces to explain its final extinction.

We know that the eastern American forests were cut down and that old nut-producing trees were probably preferentially removed, but it is difficult to imagine that the resources for the Passenger Pigeon were reduced by more than 99%, and if this species sometimes had reached 10 billion birds then a 99% reduction in resource level should leave us with a much-depleted but still immense population of 100 million Passenger Pigeons. Let's say that resources were reduced by 99.9% – we might still expect 10 million Passenger Pigeons to be flying around the skies of the USA.

Although I believe that the deforestation of large parts of the Passenger Pigeon's range was far and away the main cause of its population decline, it doesn't seem likely to have been the last nail in its coffin.

Industrialised hunting such as that at Petoskey surely played a big role in finishing it off, although it seems that most hunters gave up in the years after the great Petoskey massacre of 1878. This sort of disruptive uncontrolled and unregulated hunting was surely pushing the Passenger Pigeon closer and closer to extinction, and particularly so because its impact increased as the population decreased. As the population fell during the 1850s, 1860s and 1870s the impact of the hunting at colonies probably increased dramatically in proportional terms, with the number of hunters, their speed of arrival, their ability to catch and kill birds and the size of the markets they could supply all increasing. But was it the final nail or the penultimate nail in the coffin?

Some have suggested that Passenger Pigeons simply 'gave up' and stopped trying as their population level fell very low. This argument invokes a form of the Allee effect. Warder Clyde Allee (1885–1955) was born in Bloomingdale, Indiana. His name is attached to a possible effect where, for some species, as they decline in numbers, life becomes tougher rather than easier – this is just the type of effect which could explain the final demise of the Passenger Pigeon. But suggesting that the Passenger Pigeon simply moped its way to extinction is a bit far-fetched for me. When did the moping start – at a mere million remaining birds, or a mere hundred thousand birds? No – I don't buy it and feel pretty confident in deriding it!

Proponents of this 'final nail' refer to the colonial nesting of the Passenger Pigeon and suggest that individuals needed the stimulation of lots of other pigeons to get in the right mood (or to stimulate the right hormonal balance) for mating and reproduction. It's a possibility, and at least it is linked to the biology of the species, but ours is a social species too, and I'm willing to bet that if a few people were cast away on a desert island they wouldn't give up reproduction because they didn't feel like it any more. But on a more serious note, we know that Passenger Pigeons could be bred in captivity – indeed, Martha herself was a captive-bred bird – and these captive-breeding pigeons certainly hadn't needed the stimulation of millions of conspecifics.

Although the 'moping' effect really doesn't hold water we are looking for a similar type of Allee effect or some overwhelmingly strong mortality factor to explain that final rapid slide into extinction. And something linked to the social behaviour and ecology of the Passenger Pigeon would be particularly satisfying as an explanation.

I don't have a candidate for the overwhelmingly strong mortality factor, but I do have a contender which I am sure played a part, and it is to do with men (or women, but mostly men) with guns. Between 1850 and 1900, the period of the demise of the Passenger Pigeon, the human population of the USA rose from 23 million to 76 million.

At the beginning of this period there would have been some parts of the pigeon's range that were sparsely populated by people, but as time went on every part of its range would have had a larger and larger human population. The three-fold increase in numbers would have led to at least a three-fold increase in people letting off shots at Passenger Pigeons at a time when the Passenger Pigeon population was falling anyway, and so its impact would have grown through the period. Improvements in the manufacture of guns leading to greater range and more rapid fire would have added to this impact. Although the commercial hunting for distant markets became uneconomic as the pigeon population declined, it was always worth a man with a gun, usually in a field, pointing it skyward as a flock of Passenger Pigeons flew over – and there were more and more of those people as the Passenger Pigeon declined in numbers.

Indeed, in those areas where the Passenger Pigeon survived to breed last, for example in Michigan and Wisconsin, the rise of the human population was even more rapid. In both states the population increased six-fold in the period 1850–1900, so as the pigeons travelled to nest in diminishing numbers they were shot out of the air in increasing numbers. This was a time before there were limits on hunting, and long before such limits could be properly enforced. Although some of the human population rise would have been urban rather than rural – in fact, the cities would have grown more quickly – Passenger Pigeons overflew cities and we know that they were shot from rooftops and streets. The increasing numbers of men with guns would have been just the type of factor that would push a species harder and faster towards extinction.

But increasing human population is not an Allee effect. Can we find an Allee effect? I think we can, although it is, I have to say, partly good biological sense and partly conjecture.

The Passenger Pigeon was a tasty dish for humans but also for a range of medium-sized predators. Its eggs would have been tempting to nest predators such as squirrels, crows and jays. Its plump squabs would have been attractive to a range of crows and hawks, and its muscle-toned adults would have made great meals for birds of prey big enough and skilful enough to catch them – especially Goshawks, Cooper's Hawks and Peregrines. In addition to being tasty, Passenger Pigeons were sitting ducks, or doves. Unlike most colonial-nesting birds, they did not use inaccessible nesting sites such as cliffs or wetlands – they simply nested in trees which were safe to a certain extent from ground-living mammals but not from anything that could climb or fly. Moreover, unlike many colonial species such as ground-nesting terns or waders, Passenger Pigeons did not communally defend their colony from predators.

No, the Passenger Pigeon, when nesting, was an easy target, and it relied on predator dilution to get away with being a tasty pacifist nesting in easily accessible locations. This became a less and less viable strategy as the pigeon population fell.

Although large predators, both avian and mammalian, such as eagles, bears and wolves, were extirpated from many parts of the Passenger Pigeon range there is no evidence that medium-sized predators were so successfully killed. Indeed, we might expect that there was some element of what is called meso-predator release – where smaller predators benefit from the absence of the large predators and themselves become commoner. The same number (and maybe even a larger number) of medium-sized predators living in the locality of a Passenger Pigeon colony would be able to take at least the same absolute number of pigeons from the population, but given a smaller pigeon population the proportional impact of this predation would be much greater. If the local predators could take a million Passenger Pigeons, one way or another, from a colony of 100 million birds there really isn't any reason why they shouldn't be able to take a million from a colony of 10 million birds. And the same line of reasoning would apply to the winter roosts as well as the breeding colonies – a smaller Passenger Pigeon population sitting in an undiminished sea of predators would suffer more in relative terms.

I wonder whether Passenger Pigeons, when feeding away from their roosts or colonies, spent much time scanning for predators. All descriptions of feeding flocks suggest constant feeding and movement through the forests. It may well be, in fact I think it likely, that they relied again on predator dilution rather than predator avoidance. It was more important for a Passenger Pigeon to be feeding as quickly as possible for itself and for its squab back in the nest than it was to keep a lookout for predators. Yes, there were predators, but the chance that any individual would be killed was very small and gathering food quickly was the priority.

The fragmentation of the forest cover may have meant that Passenger Pigeons were forced to nest in smaller colonies, and roost in smaller roosts, and feed in smaller flocks, all of which would have lessened the species' ability to cope with a level of natural predation that it would have survived in the days of billions.

There are many examples of species that have been driven to extinction by changes in predator numbers and identity because they had evolved in particular circumstances. The most obvious, by far, are flightless species which just could not cope with introduced ground predators. The Dodo (which itself was a type of pigeon) did not recognize pigs, monkeys or hungry sailors as much of a threat. In a world without these foes the Dodo had life sorted – it was well-adapted to its niche on Mauritius – but when the hungry sailor became a feature of the Mauritius ecology the Dodo was doomed. The Passenger Pigeon had always lived in a world with predators, that wasn't its downfall, but its wholly gregarious lifestyle, which delivered protection from predation through the dilution effect, was less well adapted to a landscape of fragmented forests. As population size dropped it was caught in a 'predator trap' from which it could not escape – the impacts of predators were no longer sufficiently diluted. The Passenger Pigeon's reliance on predator dilution enabled its final extinction once its population was greatly reduced. This is a strong contender for an Allee effect.

At the time of the European invasion of North America the Passenger Pigeon was the commonest bird not only on that continent but in the world. It was far commoner than its fairly close relative the Mourning Dove. But now the Mourning Dove lives on as a familiar sight across most of the USA and the Passenger Pigeon is completely gone. The Mourning Dove eats seeds, nests in trees and is shot these days in its millions. It is, in many ways, a similar species to its extinct relative. Why does it survive in the land of the Passenger Pigeon's extinction?

I'm sure if we put our minds to it we could drive the Mourning Dove to extinction, but that would be more of a challenge because of its biology. It is a generalist feeder and thrives in the current US landscape. It nests in a dispersed fashion and can produce several broods a year through feeding on different seeds as they become available. Many nests are predated, but if the population falls then the remaining birds are left the best nesting sites and more food.

In contrast, the Passenger Pigeon was highly adapted to the native forest of eastern North America. It depended on a few types of food and its whole ecology was adapted to finding them and exploiting them where they were superabundant. Its food was depleting throughout its nesting season and so there was a need to nest quickly before the food disappeared, and this urgency was heightened by living in enormous colonies. The Passenger Pigeon needed to nest successfully more than once a year and it needed to find a second place to nest after its first nesting attempt. It was very successful in numerical terms until we intervened and disrupted its habitat. The Passenger Pigeon's biology was unsuited to the different environmental conditions which we imposed on it and its population fell. To add injury to insult we then killed it in enormous numbers, hastening its demise. In the end it may have been a combination of factors that finished it off: on the one hand shooting and trapping, and on the other its inability, in its reduced and fragmented population, to cope with the natural predators that it had shrugged off a century earlier.

The Passenger Pigeon's adaptations, the strengths that allowed it to be the commonest bird on Earth, were also its weaknesses when we disrupted its habitat.

COULD THE PASSENGER PIGEON HAVE BEEN SAVED?

Yes, it could have been saved – and it seems a bit odd that it wasn't, or at least that more effort wasn't put into saving it.

There were three main actions that would have saved the Passenger Pigeon: captive breeding (which would at least have maintained a captive population that potentially could have been used for a reintroduction project at some future time), regulating the shooting of birds, and habitat protection.

There were several reasons for the inactivity. First, it wasn't at all clear in the late nineteenth century that the Passenger Pigeon was heading for extinction, though over a century later one benefit of hindsight reveals that the signs were there. The species was synonymous with abundance, and its nomadic lifestyle meant that it was always a species that came and went – so as it was in fact on the verge of extinction people naturally thought that this was just a longer than average wait for a return to abundance. The biology of the Passenger Pigeon made it a species whose decline was less likely to be noticed than most.

And if it were noticed, as it was by some, whose job was it to save the pigeon? There was no US Fish and Wildlife Service in existence and no powerful NGOs with a remit to save species. Such bodies emerged in the last years of the Passenger Pigeons' existence, but by the time of Martha's death they were still only fledgling organisations with few resources.

It is particularly surprising that a captive-bred population wasn't maintained and that Martha wasn't used to keep the line going. She had herself been bred in captivity, and it seems as though captive-breeding was indeed an option – but nobody really put their minds to it.

Attempts were made to stop the carnage of shooting at nesting colonies. Laws were passed in several states, but their enforcement was lax and indeed there is a celebrated case of a team of law enforcers sent to Petoskey to regulate shooting who instead joined in the slaughter. In the woods of Petoskey, spread over scores of miles, there were armed men shooting, trapping adult Passenger Pigeons and dislodging squabs from their nests – this would have been a difficult activity for a few lawmen to control, and those asked to do it may not have been highly motivated by the task. But, generally speaking, shooting and trapping were largely unregulated in the USA until legislation of the last few years of the nineteenth century and the early years of the twentieth.

Habitat protection was on the brink of being implemented, and indeed the first US National Park came into existence in 1872, but by the time that the idea really took off apace, in the Theodore Roosevelt presidency, it was too late for the Passenger Pigeon. Look at the map of current National Parks in the USA: they contain some fantastic areas for wildlife, they are of a scale that Europeans can only envy – but only one, the Great Smoky Mountains, dedicated in 1934, would have had any relevance to the Passenger Pigeon. This great American idea of National Parks came along just a few decades too late to save the species.

More generally, the growth of environmental awareness in the USA, which burgeoned at the turn of the nineteenth century, came too late for the Passenger Pigeon. The speed with which the 'Blue Meteor' dived to extinction matched its speed through its native woods and skies – if it had tarried longer, then maybe the USA could have done more for it. If the Passenger Pigeon had survived in the wild into the early years of the twentieth century, even in low numbers, then maybe it would still be with us, gracing the skies and darkening the sun.

OTHER AVIAN EXTINCTIONS

The Passenger Pigeon was not unusual amongst pigeons in being driven to extinction by humans. A number of other pigeon species had gone before it. The most famous of these was the Dodo (extinct in around 1662), which was a rather unusual flightless pigeon. The other species were the Reunion Pigeon (1700s), the Rodrigues Solitaire (1745), the Rodrigues Pigeon (before 1750), the Tanna Ground-dove (1774), the Liverpool Pigeon (sometime around 1800), the Norfolk Island Ground-dove (c.1790) and the Mauritius Blue Pigeon (1830s) – all of these preceded the Passenger Pigeon to oblivion. The timing of the Passenger Pigeon's extinction coincided with that of a much rarer and more localised species, the Bonin Wood Pigeon from the Ogasawara Islands off Japan, which was first seen in 1827 and last seen in 1889 – it is thought to have suffered the triple-whammy of deforestation, introduced predators (rats and cats) and hunting. Later the Choiseul Pigeon was discovered by Europeans in 1904 but not reliably seen thereafter. Further extinctions of pigeons were the Red-moustached Fruit-dove (1922), the Thick-billed Ground-dove (c.1927) and the Ryukyu Pigeon in 1936. All of these species lived on relatively small islands where they were vulnerable to local events (felling of forests, introduction of non-native predators) and the Dodo and Rodrigues Solitaire were flightless too. If you were trying to make them extinct, none of them would have presented the same kind of challenge as the Passenger Pigeon. Each was an easy target compared with a widespread and abundant North American bird. And these 14 extinct pigeons form more than 10% of the 130 birds that have been driven to extinction since around 1500. Pigeons have suffered more than most groups of birds – who said the meek shall inherit the Earth?

Of these 130 extinct birds, most (85%) have lived (and died) on oceanic islands, and only around 19 have been continental species. The usual CV for an extinct bird includes terms such as 'flightless', 'island-dwelling' and 'restricted range' – none of which applies to the Passenger Pigeon. And the loss of the Passenger Pigeon from the Earth removed more individual birds than did all the other 129 extinctions put together. By any measure, this was an exceptional extinction.

We use the phrase 'As dead as a Dodo' to express something that is long gone and cannot possibly return. The alliteration between 'dead' and 'Dodo' is as pleasing as the loss of this bird is displeasing, but maybe we should replace that phrase with 'as passed as a Passenger Pigeon' – because the loss of this species was a much more exceptional event and the Passenger Pigeon is just as surely lost from Earth. As Oscar Wilde might have said, to lose the Dodo may be regarded as a misfortune but to cause the extinction of the Passenger Pigeon, once the commonest bird on Earth, looks like a misfortune which it is difficult either to understand or forgive.

WHICH HORSEMAN SHOULD WE BLAME?

There is no mystery about why the Passenger Pigeon declined so dramatically from billions of birds to millions over a few centuries, and very precipitately in the second half of the nineteenth century. But nor is there one single reason for this decline unless we call that reason 'Progress', for this once abundant species suffered numerous assaults on its habitat and on itself.

The main factor in its decline was undoubtedly the removal of its forests – it was a forest-dwelling bird that specialised on the mast of the forest like no other species. As we felled its forests and removed the older trees that produced the most mast, so the carrying capacity of its range plummeted, and with that so did the population of the Passenger Pigeon. It needed to nest successfully more than once a year whenever conditions allowed, and it needed massive blocks of forest distributed over its range to provide for this. Large-scale deforestation removed the options for Passenger Pigeons to nest often enough in the same year, thus reducing annual productivity to below a critical threshold.

The rise of commercialised killing of Passenger Pigeons for distant markets was a disaster for this bird and hastened its demise. This killing had all the hallmarks of unsustainable harvesting – it involved a large number of unregulated and competing interests, it was carried out in the breeding season, it involved indiscriminate killing of adults and young and was carried out in ways that were bound to lead to huge disturbance of the unculled birds, thus reducing their nesting success as well. This assault will have reduced numbers through direct losses and through greatly reducing the number of young reared in affected colonies.

Although the last few decades of the Passenger Pigeon encompassed the spread of introduced European birds, such as the House Sparrow, there is no strong evidence that introduced diseases played an important role in its demise – although had it survived another 20 years then it would have experienced another assault on its habitat as the native American Chestnut disappeared from the landscape due to an introduced disease of trees.

The insidious effect of an increased human population through opportunistic and unregulated shooting throughout its range probably increased in severity and impact as time went on, and perhaps as the population declined its ability to withstand the attacks of natural predators became lessened year on year so that they too played an increasing part in driving the bird to extinction.

The Passenger Pigeon was not killed by a single horseman of the ecological apocalypse – it suffered multiple wounds. If we were to look at the 300 years of decline of the species and apportion 'blame' to different factors for its extinction, then I would allocate the percentages as follows. Chains of extinction played no material part (0%) though, ironically, had the Passenger Pigeon survived a few more decades then the loss of the American Chestnut might have had a significant impact. I harbour a sneaking suspicion of the role of exotic diseases in the story, not as a primary cause, but perhaps as a contributory one – and wouldn't it be highly ironic if it were the spread of one European invader, the House Sparrow, through the actions of another, ourselves, that spread diseases that might have played a role in this story? Pigs gobbling their way through some of the Passenger Pigeon's food supply may have played a part, but I think we can only give competitive pigs and exotic diseases a combined score of 4% under the heading of Introduced Species.

The major factor was habitat destruction, and I would allocate 85% of the 'blame' to that factor in its different manifestations – if you cut down the forests then a forest-dwelling species is going to suffer a lot. In its weakened state then overexploitation, by market hunters but also by casual shooting, trapping and collecting of squabs, and the disturbance that these activities caused to breeding and nesting sites, contributed, in my view, 10% of the total. That leaves 1% – and that last 1% was the Passenger Pigeon's biology. Its strength, as with many of us, was also its weakness. Its ability to be a tasty, defenceless dove depended on its strength in numbers – and once those numbers were greatly reduced by other factors then its final demise was hastened by its inability to withstand the impacts of natural predation. And Buttons was shot, and Martha died of old age. That is the tale of the Passenger Pigeon's extinction.

Really, it was progress and civilisation that drove the Passenger Pigeon to extinction. The human population, swelled by immigrants from Europe, was increasing every year, and the people moved west each year, pushing back the frontier, increasing the demands on the land and water and requiring ever more resources as the population grew. People wanted timber for railways, for housing, for firewood and also because they wanted the land for farming. They wanted cheap meat too. They wanted House Sparrows and pigs. It was we, the European invaders (I write as a European), who drove the Passenger Pigeon to extinction. We didn't intend to do it, but we all played a part. There's no point blaming the man with a gun in Wisconsin – as he was supplying a woman in New York City, Boston or Chicago. Was it the man or the woman who was to blame? Or was it what we like to call Progress?

The uncomfortable thing about all this is that nobody wanted to drive the Passenger Pigeon to extinction, but we did. And we did it through building railroads, inventing better guns, increasing the human population of the USA, getting rid of the Native Americans who had lived in some sort of balance with the pigeons for thousands of years, and generally bringing the trappings of civilisation to the continent. It was all the things that we hold dear that did for the Passenger Pigeon.

If we want to understand a little more about who killed the Passenger Pigeon, then we must meet another Martha, a woman named Martha Grier (nee Dean). The next chapter tells her story.

CHAPTER SIX

How the Wild was lost when the West was won

Alas, the pigeons and the frosty morning hunts and the delectable pigeon-pie are gone, no more to return. They are numbered with those recollections which help to convince me that the boys of to-day don't have as good times we youngsters did in the prime of our busy out-door world.

W. B. *Mershon* (My Boyhood Among the Pigeons, *1907*)

On the day that Martha the Passenger Pigeon died in Cincinnati, Ohio, another Martha died in the town of Bridgeport, in the east of the state. Martha Grier was the only woman named Martha to die in Ohio on that day, and according to her death certificate she passed away at 9:45 am, aged 76 years, seven months and 24 days. Approximately three hours later, between midday and 1 pm, the last Passenger Pigeon expired at an age of somewhere between 17 and 29 years.

Both Marthas died of old age, both had spent most of their lives in Ohio, and in neither case was the death unexpected. The *Cincinnati Enquirer* wrote on 18 August that 'the days of the last Passenger Pigeon ... are numbered', as Martha had been listless and almost motionless for some time. Martha Grier had been attended by her doctor since 15 August; she died from paralysis and senile decay.

The two old Marthas faded away on the same day, in the same state, and each from her own separate version of old age. They shared one further link – they were both widows whose partners were named George. The pigeons had been named in honour of George Washington and his wife Martha, and George, the last male Passenger Pigeon, had died in July 1910. His companion was thus left as the last surviving member of her species, named after the USA's first First Lady but now, ironically, the Passenger Pigeon's Last Lady. Martha Grier had married George Ross Grier some time in the early 1860s; he died in January 1904, 10 years before Martha herself.

The long-expected death of the last Passenger Pigeon attracted some newspaper headlines. The *Cincinnati Enquirer* noted that there would be no funeral for Martha but that her body would be shipped to the Smithsonian Institution in Washington to be put on display, and that is indeed what happened. The last Passenger Pigeon was shipped east in ice, as had been many of her predecessors, but this time not packed in a barrel or crate with scores of others but alone in a 140-kilogram block of ice, and not to be feasted on in a restaurant or dining room, but to be viewed in a glass case.

The Daybook of Chicago, Illinois, on 2 September also mentioned the passing of Martha and that there had been millions of wild Passenger Pigeons still in existence when she had hatched. A reward of several thousand dollars was offered for any information leading to the capture of a pair. In New York, *The Evening World* of 7 September covered the extinction with the thought that the last Passenger Pigeon's death should be remarked on even in those days when 'the last European soldier known to war lords is in danger of dying in his turn' – for these were the early days of the First World War. The newspaper went on to mention Wilson's two-billion flock sighting, and that the Bishop of Montreal (in 1703) excommunicated the pigeons because of the damage that they did to crops. The piece ended on this note: 'Therefore, while we may lament in sentiment the passing of the last Passenger Pigeon, we have good cause to rejoice that nature did not fit him to adapt to civilisation and stay with us like the grasshopper.' Martha Grier's passing was less noticed.

Martha Grier's life was that of an ordinary person in the American story. Her life encompassed the rapid growth and expansion of the American nation, its Civil War, the abolition of slavery and major economic and social changes. It also encompassed the events leading to the extinction of the Passenger Pigeon: the loss of forests, the growth of the human population and the mass killing of pigeons for distant markets. The Passenger Pigeon was not alone in being driven to extinction in this period – there were several other extinctions and near-misses for other species. Martha Grier lived through the period when Progress marched across the continent and when the Wild West was tamed – when the West was won, the Wild was lost.

By examining the events of US history that occurred through Martha Grier's life we will come to a better understanding of the USA that drove the Passenger Pigeon to extinction. By placing that extinction in its wider context, we will also come to see how it stands as a totemic example of environmental loss during a period of rapid social and economic change.

1838

Martha Grier was born Martha Dean on 8 January 1838 to immigrant parents (Scottish father, English mother), in Coshocton County, Ohio – just 75 miles (120 km) from where she died.

Elk were extirpated from Ohio.

26 US states in existence – the original 13 that ratified the Constitution (Delaware, Pennsylvania, New Jersey, Georgia, Connecticut, Massachusetts, Maryland, South Carolina, New Hampshire, Virginia, New York, North Carolina and Rhode Island) and (in order of accession) Vermont, Kentucky, Tennessee, Ohio, Louisiana, Indiana, Mississippi, Illinois, Alabama, Maine, Missouri, Arkansas and Michigan.

The 87th and last set of plates of **John James Audubon's** *The Birds of America* was issued. The plates were unbound and were without text so as to avoid having to furnish free copies of the 'book' to some libraries.

Cherokee Trail of Tears – the Cherokee Native Americans from southeastern states, such as North Carolina and Georgia, were forcibly moved into concentration camps and then marched to the area that much later became the state of Oklahoma. *En route* thousands died because of the harsh winter conditions. The Cherokee were the last of the 'Five Civilised Tribes' to be treated in this way – the others were the Choctaw, Creek, Seminole and Chickosaw. In the 1820s about 120,000 Native Americans lived east of the Mississippi, but by 1840 the number was only 40,000.

John Muir was born in Dunbar, Scotland. Muir moved to the USA at the age of 11 and lived on a farm near Portage, Wisconsin. He played a central role in advocating the preservation of the American wilderness through National Parks. Not only was he born in the same year as Martha Grier, he also died in the year of her death.

1839

The first state law **allowing women to own property** was passed in Mississippi.

1840

The **census** enumerated the American population (not counting Native Americans) as 17,069,453 (around one in seven was a slave). Ohio had a population of 1,519,467, of whom 21,590, including the two-year-old Martha Dean, were in Coshocton County. The UK population (1841) was around 26 million.

1841

President William Henry Harrison died a month after taking office. He died of pneumonia, perhaps facilitated by giving the longest ever inauguration speech, of two hours, on a freezing cold day. He was elected from Martha Dean's home state of Ohio. Harrison was the first of six US presidents with Ohio as their state of primary affiliation, and all but one of them served during Martha Grier's lifetime.

1842

John Charles Fremont's first expedition. Known as the Great Pathfinder, Fremont was one of those who opened up the West, mapped it and encouraged its exploration and settlement.

His first expedition was to the Continental Divide in Wyoming. Later expeditions took him to California, where he became rich in the Gold Rush. He was the Republican Party's first presidential candidate in 1856 ('Free soil, free men, and Fremont!'), and for a while, before coming to a deal with Abraham Lincoln, was presidential candidate for the Radical Republicans in 1864. He was an active, though not always prudent, Union general in the Civil War.

1843

The **Wagon Train of 1843** consisted of around 1,000 people, and their wagons, who followed the Oregon Trail from Missouri to Oregon. More than 400,000 people used the trail to travel west until the completion of a transcontinental railway (1869).

1844

Samuel Morse sent the first telegraph message from Washington to Baltimore.

1845

Baseball codified. How can you take seriously a sport which is based on the Knickerbocker Rules?

Florida and Texas became the 27th and 28th states, respectively. The US states share sovereignty with the US government, hold elections and have their own governments. Prior to becoming states areas governed by the USA were territories. Territories became states once they had sufficient population and the social and legal infrastructure to play a full part in US governance.

The Spanish conquistador Juan Ponce de León named Florida the 'flowery land' in 1513 – he is said to have met a native tribesman who already spoke Spanish. Florida was fought over and traded in treaties by Britain, France and Spain, but in 1813 Spain ceded Florida to the USA at a price of $5 million and for a promise that the USA would renounce any claims to Texas. After Andrew Jackson's Indian Removal Act of 1830 the native inhabitants of Florida fought three wars against the European invaders and the US government spent tens of millions of dollars fighting back. Thousands of Seminole Native Americans from the swamplands of the Everglades were moved to the Great Plains Indian Territory to the west of the Mississippi River – a completely different habitat in which their customs and ways of life were foreign, in order for the Europeans to settle and develop the land.

Texas, the Lone Star State, was colonised by French invaders in the late seventeenth century but was largely under Spanish control until Mexican independence from Spain in 1821. Texas declared independence from Mexico and became a republic in 1836. Even then, some were seeking closer links with the USA, and nine years later Texas joined the Union.

1846

Iowa became the 29th state. The first settlers officially moved to this area in 1833, and the Iowa Territory was established in 1836. By 1840 the population was around 43,000, but by 1850 it had risen to 192,000. This was an area of tall-grass prairie when settled, but only 1% of that original habitat remains.

Start of **US–Mexican War**.

The **US and British governments agreed to divide the disputed 'Oregon country'** along the 49th parallel, with British Columbia to the north, in what is now Canada, and what are now the states of Washington and Oregon to the south.

1847

Samuel Colt sold a revolver pistol to the US government. Colt's made a fortune in the Civil War, and Samuel Colt was the USA's first manufacturing magnate.

The **USA issued its first postage stamps**. The 5-cent stamp bore an image of Benjamin Franklin (the first US Postmaster General) and would pay for delivery of a letter up to 300 miles (483 km) distant; the 10-cent stamp, George Washington, would take letters further.

1848

The **Grey Wolf** was extirpated in Ohio.

Wisconsin became the 30th state. Its population had been over 3,000 in 1830, more than 30,000 in 1840 and now, after statehood, rose to more than 300,000 in 1850.

The **Mexican Cession** after the defeat of Mexico in the US–Mexican War led to Upper California and areas of Arizona coming under US control.

1849

The **California gold rush** (1848–55) led to hundreds of thousands of European invaders entering California from the sea and overland. The search for gold led to the deaths and displacement of tens of thousands of Native Americans who were murdered or chased off their ancestral land.

1850

The **census** suggested that the population was 23,191,876 – more than one in eight of whom were slaves and around one in ten of whom were foreign-born.

California was admitted to the USA as the 31st state and the first on the Pacific coast. The first European involvement with California was through the Spanish as part of their empire covering Florida and much of the southwestern part of the USA. Mexico became independent from Spain in 1821 and the US–Mexican War of 1846–48 led to California becoming part of the USA. It is now the third-largest state in area (after Alaska and Texas) and home to one in eight Americans.

The **Mountain Lion, Canadian Lynx, Fisher and American Marten**, four mammalian carnivores, were all extirpated from Ohio by this year.

US annual lumber production stood at a paltry five million board feet.

1851

In New York City, John James Audubon died, the *New York Times* was first published, and **eight pairs of House Sparrows were introduced to Brooklyn**.

Within 50 years House Sparrows were established in all US states. Cincinnati was another centre where birds were released, and Martha (the Passenger Pigeon) must have seen House Sparrows hopping around in freedom while she spent her years in captivity. I wonder when Martha Grier saw her first House Sparrow – a species that would have been familiar to her British parents but novel for her. House Sparrows were introduced in the hope that they would control insect pests but they became a pest themselves, as introduced species so often do – for they were, as any European farmer would know, also partial to cereal seeds. 'Without question the most deplorable event in the history of American ornithology was the introduction of the English Sparrow' – W. L. Dawson, *The Birds of Ohio*, 1903.

Black-footed Ferret discovered – or at least its existence was publicised in Audubon's and Bachman's book *Viviparous Quadrupeds of North America*. Native American tribes in its original range of southern Canada through to Arizona knew it well and used it in rituals and for food. Black-footed Ferrets prey on prairie dogs, rodents whose enormous underground colonies used to cover the short-grass and mixed-grass prairies. With the conversion of the prairies to croplands, declines in prairie dog 'townships', the arrival of canine distemper and being trapped for fur, the Black-footed Ferret declined and was declared extinct in 1979. Reports of its extinction, like those of Mark Twain's death in 1879, were an exaggeration, as in 1981 Lucille Hogg's dog Shep brought home a dead Black-footed Ferret. A small population was located in Wyoming, which then disappeared by 1987 largely because of introduced diseases. Now a reintroduction project has established a population of around 1,000 Black-footed Ferrets through reintroductions in Canada, Wyoming, Montana, South Dakota, Utah, Kansas, Arizona and Mexico. The prospects now look quite good for this species to survive, albeit after huge historical losses and the expenditure of millions of dollars.

Herman Melville published ***Moby-Dick***, a novel in which nature fights back.

1852

Last Great Auk sighting off the fishing grounds of the Grand Banks off Newfoundland, Canada. This large flightless auk lived on fish such as Capelin, and also crustaceans. It was a strong swimmer but very vulnerable on land. It was hunted for its down, and towards the end of its existence its rarity increased the avidity with which museums sought specimens.

Xerces Blue butterfly first described. The Xerces Blue was a butterfly of restricted range – it lived in coastal sand dunes of California. Its problem was that it lived in those coastal dunes that are now the Sunset, Seacliff and Richmond districts of San Francisco. Housing development put an end to this species – the butterflies didn't cope well with concrete replacing their sand-dune homes. The introduction of a South American species of ant, which displaced native ant species, may have had a role to play too, as many blue butterflies have evolved symbiotic relationships with native ant species. The Xerces Blue is thought to be the only US butterfly to be driven extinct in historic times, and its end came some time between 1936 and 1943, less than a century after we realised that we shared the planet with it.

Harriet Beecher Stowe's ***Uncle Tom's Cabin*** (an anti-slavery novel) was published, and sold more than a million copies in the year.

1853

Rumour has it that the **potato chip was invented** by Native American cook, George Crum, to pacify a disgruntled customer in Saratoga Springs, New York.

1854

The **Republican Party was founded** in Ripon, Wisconsin, as an anti-slavery party. The Democrats had already been around for about 26 years.

The **Gadsden Purchase** (from Mexico) was ratified by the US Senate, adding almost 77,000 square kilometres of southern parts of what are now Arizona and New Mexico to the USA. This block of land completed the extent of the current contiguous 48 US states (though they did not all become states until 1912).

1855
A quiet year in US history.

1856
Woodrow Wilson, who would be president when the two Marthas died, was born in Staunton, Virginia.

1857
Galen Clark and Milton Mann were the **first European Americans to see the Mariposa Grove of Giant Sequoias** in Mariposa County, California. The oldest tree in the grove, Grizzly Giant, is more than 2,000 years old.

1858
Minnesota became the 32nd state.
Theodore Roosevelt, future president, was born in New York, New York.

1859
Oregon became the 33rd state.

1860
Martha Dean was living in the house of the grocer Jacob Shelling and his wife, Mary, and their children Eugene (12 years), Marriett (10 years) and Anna (four months) in Goshen in Tuscowaras County, which neighbours her own Coshocton County. Perhaps Martha was helping to look after the children – but she was more likely to have been a shop assistant.
George Ross Grier, Martha's future husband, was living in the Braggs household in Bridgeport, Belmont County, Ohio, with R. Braggs (master carpenter) and his wife Margaret and six offspring, a domestic servant Mary Weaver and her daughter

Alice. Bridgeport was where George and Martha would live and bring up a family of their own, but the Civil War would first intervene.

The **census** enumerated the population of the USA at 31,443,321 – one in eight of whom were slaves.

The **repeating rifle** was invented.

The **Pony Express** operated a mail service from St Louis, Missouri, to Sacramento, California, for 18 months until the telegraph made it redundant. Riders, using a range of relay stations, covered the 1,900-mile (3,000-kilometre) distance in 10 days. The most famous Pony Express rider was 'Buffalo Bill', William Cody, who signed up as 15-year-old, perhaps in response to the advertisements that read 'Wanted: Young, skinny, wiry fellows not over eighteen. Must be expert riders, willing to risk death daily. Orphans preferred.'

US annual lumber production was a modest eight million board feet.

1861

Abraham Lincoln became the first Republican President. Lincoln was born in Kentucky (south of Louisville), brought up in Indiana and elected by Illinois.

The **American Civil War** began.

The first **cross-continental telegraph** was completed.

Kansas became the 34th state. Kansas was part of the land bought by the USA from France in 1803 as part of the Louisiana Purchase, which included all of the current-day states of Arkansas, Missouri, Iowa, Nebraska, Kansas and Oklahoma; large parts of Minnesota, South Dakota, North Dakota, Montana, Wyoming, Colorado and Louisiana; and smaller portions of New Mexico and Texas. They were purchased for $11.25 million (plus the cancelling of debts of $3.75 million) and covered 2,140,000 square kilometres. This seems quite good value for about a quarter of the current USA. The world would have been a very different place had France held on to this land.

As Kansas was officially opened up in 1854 there were violent clashes, described as 'Bleeding Kansas', over whether it would be a Free State or a Slave State. At one stage, Kansas had two governments, one pro- and one anti-slavery. Kansas entered the USA as a Free State.

Kansas and much of the other land bought from France was grassland comprising the Great Plains. Three types of grassland are recognised and they sound like the coffee options in a Starbucks café: Tall-Grass Prairie (the easternmost flavour), Mixed-Grass Prairie (the one in the middle) and the western Short-Grass Prairie. Kansas had all three types and clings on to remnant areas of all three too.

1862

George Grier, soon to be Martha Dean's husband, **enlisted on the Union side** in Company B, 99th Ohio Volunteer Infantry, on 15 August with the rank of sergeant. Ohio was a Union State, but across the Ohio River from Bridgeport, where George was living and working at the outset of the War, was the Confederate State of Virginia.

The 99th OVI was mustered from several western counties of Ohio, far from Bridgeport where George had been living in 1860. Perhaps this was the part of Ohio where George was born, and perhaps he returned there to enlist with friends and relatives. The 99th OVI was sent to Kentucky and spent December near Nashville, Tennessee.

The first **Homestead Act** was passed after southern states seceded from the Union. Up until this point they had vetoed moves to provide land grants to individuals in the West, fearing that it would lead to an increase in European immigration. The Republican Party favoured the democratisation of land ownership in accord with Thomas Jefferson's model of the USA as being built on individual self-sufficient farmers. Land was allocated to applicants who had not taken up arms against the US government, who were older than 21 and who could meet residency conditions, and it was allotted in quarter-sections of 160 acres (65 ha).

William Sullivant watched a flock of 25–30 **Carolina Parakeets**
fly over Columbus, Ohio – the last known Ohio record.

1863

George Grier and comrades spent the year moving south through
Tennessee to the area around Chattanooga.

An autumn flock of **Eskimo Curlews** in Nantucket was said to
have 'darkened the sun' and exhausted the island's ammunition
as 7,000–8,000 birds were shot from this multitude.

Eskimo Curlews wintered (and maybe still do) on the pampas
of Argentina and took a northwards route through the Great
Plains to the tundra of Canada and Alaska (which at this time
was Russian territory). Their southward migration took them
down the eastern seaboard of Canada and the USA and back to
Argentina. This was one of the commonest shorebirds in North
America and may have been the species that Columbus's crew
spotted which persuaded them that they were not far from land
(which is somewhat ironic, as this bird is thought to have taken
a migration which departs far from shore).

Their alternative names of 'doughbirds' came from the fat
that they laid down in autumn, and 'prairie pigeons' showed
that they were seen as the abundant alternative to the fast-
diminishing population of Passenger Pigeons. In the late
nineteenth century upwards of two million were shot annually,
many in spring, when such killing will have the largest
population impact.

Canning factories were set up in Labrador in the 1880s, but
a decade later commercial exploitation had so reduced Eskimo
Curlew numbers that they were on the verge of extinction –
and their numbers never recovered. The last generally accepted
records were in 1963 from Galveston, Texas. Nobody is holding
their breath and expecting this species to recover – even if, as
seems a little unlikely, it is just holding on.

West Virginia became the 35th state, a Free State, by seceding
from Virginia, a Confederate State.

1864

George Grier's regiment was involved in the **siege of Atlanta, Georgia**, in the summer and the pursuit of General Hood's defeated forces into Alabama in September. In mid-December they were a part of the battle of Nashville back in Tennessee and were again chasing the defeated Hood, this time to the Tennessee River. The war was almost won, and at the end of the year Sergeant George Grier was transferred from Company B of the 99th Ohio Volunteer Infantry to Company F of the 50th Ohio Infantry.

Nevada became the 36th state.

Abraham Lincoln signed the **Yosemite Grant**, which handed over the Yosemite Valley and the Mariposa Grove of Giant Sequoia to the State of California for 'public use, resort, and recreation ... to be left inalienable for all time ' – and this paved the way for the creation of National Parks. Galen Clark was, for 24 years, the first guardian of Yosemite.

1865

On 5 March 1865, **George Grier was mustered out of the Army** as being 'supernumerary by reason of consolidation'. The war was almost over, the South was defeated. George Grier's regiment had lost 342 men during their service: 84 through combat and 258 through disease. It was time to go back to Ohio and resume a normal life.

President Lincoln was assassinated by John Wilkes Booth and succeeded by the vice-president, Andrew Johnson.

Horace Greeley, the founder of the *New York Tribune*, wrote in an editorial of 13 July: 'Washington is not a place to live in. The rents are high, the food is bad, the dust is disgusting and the morals are deplorable. **Go West, young man, go West and grow up with the country.**'

Around this time the human population of the USA reached **35 million** – which was about the same as the population of the **Pronghorn antelope** before the European invaders discovered it. One of the three major grazers of the Great

Plains, along with the Bison and prairie dogs, the Pronghorn is the sole antelope living in North America. Perhaps the fastest species on Earth, possibly faster than the Cheetah and certainly able to run fast for far longer, the Pronghorn's speed in nose-diving towards extinction was also impressive. Possibly more numerous on the Great Plains than the Bison, the Pronghorn was reduced to around 13,000 individuals (a 99.96% decline) by 1915 – since when its fortunes have improved through conservation efforts, and there are now upwards of 700,000 individuals (a mere 98% decline).

The Pronghorn evolved when there was an American species of cheetah, and the antelope's speed is thought to be an evolutionary hangover from spending millions of years running away from that – but it couldn't run, nor could it hide, from the destruction of its prairie grasslands, fencing of rangelands (they can run but they can't jump, having evolved in a habitat where there wasn't much to jump over) or overhunting.

Sharecropping replaced slave plantation during the reconstruction phase following the Civil War. Freed slaves could work on the land and be paid by their former 'owners'. The sharecropper received accommodation, tools, seed and the land and kept between a third and a half of the crop as his 'share'.

The **Ku Klux Klan** was founded. It persisted until 1874 before being revived in 1915.

1866

Jesse James and his gang started robbing banks, and carried out the first daylight bank raid in the USA in St Louis, Missouri.

Robert Leroy Parker – later known as **Butch Cassidy** – was born in Beaver, Utah.

1867

George and Martha Grier's first daughter, Mary Lurene, was born in Belmont County, Ohio; as was Harry Alonzo Longabaugh – later known as the **Sundance Kid** – in Mont Clare, Pennsylvania.

Nebraska became the 37th state.

The **USA purchased Alaska** from Russia for $7.2 million.

Barbed wire and the **refrigerator car** were invented.

1868

Male suffrage for over-21s, which would have included George Grier, was brought in through the 14th Amendment.

1869

George and Martha Grier's **second daughter, Frances (or Fannie), was born**.

Heath Hens extirpated on the American mainland. This bird, perhaps 'just' a subspecies of the Greater Prairie Chicken, was abundant in the areas first occupied by the European invaders. Heath Hens, rather than Wild Turkeys, may have been the roast bird on the table for the first Thanksgiving Dinner of the Pilgrims in 1621. The Heath Hen lived from southern New Hampshire to Virginia, at least, and was originally the poor man's (and woman's) food as it was so cheap. Servants would sometimes seek to have their contracts stipulate that they should not be fed this meat more than two or three times a week. But from this year onwards it remained only on Martha's Vineyard, a Massachusetts island, where it increased in numbers from 300 to 2,000 before plummeting to extinction in 1932. Americans seem keener on naming the last birds of their species than they do on saving them – the last male was called 'Booming Bill' – but he didn't boom, he bombed.

The Union Pacific–Central Pacific **transcontinental railroad is completed** as the two lines meet at Promontory Point, Utah.

1870

George, Martha and family were living in Bellaire, Belmont County, Ohio, close to where George had been living in 1860. George was working as a mail agent. Martha was 'keeping house' and bringing up their two young daughters, Mary and Frances.

The **census** estimated the USA population at 38,555,983, none of whom were slaves.

US annual lumber production reached a fast-increasing 13 million board feet.

Railroad mileage reached 50,000 (80,000 km).

1871

The largest single well-documented nesting of Passenger Pigeons took place at **Kilbourn City, Wisconsin**.

American Acclimatization Society founded in New York City with the aim of introducing European fauna and flora into the USA for economic and cultural reasons. A leading light of the society was Eugene Schieffelin, who wanted to introduce all the species mentioned by Shakespeare into the USA. Most attempts at introduction failed, but the Starling (*Henry IV, Part 1*) was spectacularly successful and fairly disastrous. There are now more than 200 million Starlings across the USA, and barely a day will pass when the bird-savvy traveller fails to see them.

1872

President Ulysses S. Grant signed the law that made **Yellowstone the first National Park** in the world. It now covers nearly 9,000 square kilometres of forest, lake, mountain and geothermal springs. Most of the park is in Wyoming but it extends into Montana and Idaho. It comprises the largest near-intact ecosystem of northern temperate areas and its wildlife stars include Black and Grizzly Bears, Grey Wolves, Elk and Bison. Some of the earliest tourists to visit the park were killed by Native Americans, but now around two million people visit Yellowstone every year (the very few who die these days have usually taken liberties with nature – either with bears or with geothermal springs).

Forest cover of the eastern USA reached its lowest point in history – 48% of its extent in 1620.

1873

Levi Strauss and Jacob Davis received a US patent to make the **first riveted denim work pants**.

1874

George's and Martha's son, Charles, was born.
Cincinnati Zoo opened.
The **Reverend John Bachman died**. Bachman found a new
species of warbler in 1832 and gave specimens to Audubon,
who then named the species after his friend.

The Bachman's Warbler nested in the southeastern USA in
canebrakes and swamp forests and wintered in Cuba. Arthur T.
Wayne once shot 11 in three hours on a scientific expedition to
Florida. On his 1892 collecting trip he shot 43 Bachman's
Warblers (and six Carolina Parakeets and 13 Ivory-billed
Woodpeckers). In 1906 Wayne found a Bachman's Warbler nest
with two eggs in the I'on Swamp just north of Charleston, South
Carolina – where it was also last seen in the USA, in 1962.

The cartoon strip *Doonesbury* had a character, Dick Davenport,
who was a birder, and who had a heart attack when he saw and,
with his last action on Earth, photographed, a Bachman's
Warbler.

1875

Last record of the **Labrador Duck**. The Labrador Duck was
always fairly rare, but it wasn't always extinct. It bred in
Labrador, probably, and wintered in sandy bays, where it ate
marine shellfish, around New York and New Jersey and as far
south as Chesapeake Bay. Although it was shot it wasn't prized
for its meat, and its extinction may have been caused by over-
harvesting of its eggs or by reductions in its food supply – we
don't really know. Audubon mentioned that its flight was swift
and its wings whistled in flight.

1876

The **centenary of the US Declaration of Independence**,
which contains the words, 'We hold these truths to be self-
evident, that all men are created equal, that they are endowed
by their Creator with certain unalienable Rights, that among
these are Life, Liberty and the pursuit of Happiness.'

Colorado became the 38th state.

The **Battle of the Little Bighorn** was a victory of the Lakota Sioux and their allies over the US government forces led by Civil War hero General George Armstrong Custer. The US government had allowed, and even encouraged, gold prospectors to enter the Black Hills of South Dakota which were sacred to the Sioux. A numerically much superior Native American force, led by Crazy Horse and inspired by the visions of victory seen by Sitting Bull, overwhelmed and massacred a small US Seventh Cavalry force, which became divided during the battle, making their defeat easier. The US Cavalry lost 258 men and the Native American losses were small.

The **telephone was invented**, and Thomas Edison was one of the people who played a part in its development.

1877

Jared Potter Kirtland died – but the warbler and snake named after him lived on.

The **Nez Perce War**. The Nez Perce refused to give up their tribal lands on the borders of Idaho, Oregon and Washington states, which had been granted to them in a treaty of 1855. Around 250 warriors, and more women and children, fought a retreat towards the Canadian border, first calling upon the Crow Nation for help (which was refused), with the aim of meeting up with Sitting Bull, who had fled to Canada to escape capture after Little Bighorn. They were pursued by 2,000 US soldiers and eventually defeated in a surprise attack just 64 kilometres short of the Canadian border. The captured tribe was made to settle in Indian Territory (in present day Oklahoma) even though they had been promised a return to their tribal homeland, but then after a decade or so they were allowed back to some of the areas of the Pacific northwest where they originated.

1878

The great **Petoskey, Michigan, massacre** of Passenger Pigeons. Thomas Edison filed for the US patent for the **electric lamp**.

1879

The approximate mid-point of the **'Gilded Age' of the American economy**, which lasted from the end of the Civil War to the beginning of the twentieth century. Net national product increased at levels around 7% per annum for the previous decade and at 4.5% per annum per capita (allowing for the increase in population) and was set to continue to increase at slightly lower levels for the decade to come as well. This was the era of great industrial magnates such as John D. Rockefeller (1839–1937, founder of the Standard Oil Company), Andrew Mellon (1855–1937, banker and industrialist), Cornelius Vanderbilt (1794–1877, shipping and railroads) and Andrew Carnegie (1835–1919, steel), who made enormous fortunes and were philanthropic to boot.

The term 'Gilded Age' was a satirical one, coined by Mark Twain and Charles Dudley in their (1873) book of the same title which satirised greed and corruption in the USA in the days after the Civil War. Rather than being a Golden Age, Twain and Dudley regarded the period as having a glittery exterior which covered economic exploitation and social evils.

1880

George, Martha and family had moved a short distance up the Ohio River to Bridgeport, also in Belmont County. George was now working as a clerk in a saw mill (though on the day of the census, 8 June, he was ill at home with scarlet fever). Mary was 13, Fannie, 11, and Charles, 5.

The **census** showed a population above 50 million, of which 6.6 million were foreign-born. The UK population was 32 million.

US annual lumber production was still increasing rapidly, at 18 million board feet.

1881

Henry James published *The Portrait of a Lady* (an existentialist novel) and the Earp brothers and Doc Holliday ended the existence of Ike Clanton and Tom and Frank McLaury in the gunfight at the **OK Corral** in Tombstone, Arizona. This is perhaps a telling contrast between the East and the West at that time.

The **Black Bear was extirpated in Ohio**, although these days they are spreading back into the state from West Virginia. Do they swim the Ohio River or trot across the bridges, I wonder?

1882

The approximate **peak year in sales of Sea Otter pelts** in London – 6,500 pelts were sold.

Sea Otters are the heaviest members of the mustelid (weasel) family, weighing around 40 kilograms. They are almost entirely marine and have the thickest fur of any mammal. Sea Otter fur was prized for its warmth as clothing or bed coverings and, in China, as a trim on clothing that served as a sign of wealth.

Sea Otters live along the Pacific coast from Alaska as far south as central California (but formerly as far south as Mexico), as well as on the western Pacific shores (mostly in Russia but as far south as Japan). Their global population fell from around 200,000 to a low of around 1,500 individuals because of the fur trade. As early as the 1820s the California Sea Otter population had almost completely disappeared – now there are around 2,000 individuals, mostly in the area around Monterey. The world population is around 175,000 individuals.

1883

Standard time zones were adopted, because rail travel meant that it was possible to travel widely, quickly.

The **American Ornithologists' Union** was founded. Letters of invitation were sent out by three distinguished ornithologists, Joel Allen from the American Museum of Natural History (who was the first president of the AOU and the first editor-in-chief of its journal), Elliot Coues (who did much to establish the trinomial classification of subspecies for birds and then in zoology generally) and William Brewster (curator of mammals and birds at Harvard University).

1884

Arlie W. Schorger born in Republic, Ohio.

The **first issue of the American Ornithologists' Union's journal** was published. The journal is named after an extinct bird: *The Auk*.

The **world's first skyscraper** was built in Chicago – the Home Insurance Building, 10 storeys high.

The **low point of Bison populations**. There had once been 30 million Bison on the plains of North America, and also millions more of a woodland subspecies which was an even larger beast. Woodland Bison east of the Mississippi had been driven extinct by 1833, and 50–60 years later the population of Bison reached a low of fewer than 1,000 individuals.

At the beginning of the nineteenth century, Peter Fidler, a surveyor from the Hudson's Bay Company said that 'the ground is entirely covered by them & appears quite black. I never saw such amazing numbers together before, I am sure there was some millions in sight as no ground could be seen for them in that compleat semi-circle & extending at least 10 miles.'

The Native American Plains peoples depended largely on Bison for their food and used many parts of the beast for eating and clothing – and as the structures of their tipis. Shepard Krech, in *The Ecological Indian*, reckoned on there being 120,000 Plains Indians at the start of the nineteenth century, and that they could safely take 720,000 Bison a year and that this would meet most of their needs. It seems that the impact of Native Americans might well have been to take a sustainable harvest of Bison – or perhaps they were causing a small annual decline in numbers – but what they weren't doing was driving the species towards extinction in a matter of decades.

Railroad expansion made Bison killing and Bison transporting much easier. As with the Passenger Pigeon, the railroads brought the killers to the prey and the hides to the markets. Between 1871 and 1878 the hunters shot the southern herd to extinction and then turned their guns and attention to the north – 'little else was done' in the Kansas regions of Dodge City, Wichita and Leavenworth, 'except Buffalo killing.'

William Temple Hornaday, a conservationist and taxidermist, wrote in 1887 that the story of the Bison was a 'disgrace to the American people in general, and the Territorial, State and General Government in particular. It will cause succeeding generations to regard us as being possessed of the leading characteristics of the savage and the beast of prey – cruelty and greed.'

The Bison remained in Yellowstone National Park and a few private ranches, and it has since recovered to more than 500,000 individuals, most of which are still in Canadian and US National Parks.

1885

This may have been, at the very earliest, the year in which Martha, the last Passenger Pigeon to survive on Earth, was **hatched** in captivity.

1886

Statue of Liberty erected. The statue was a gift from France and sits on Liberty Island in New York Harbour facing towards Europe. Since 1886, it has silently, but symbolically, greeted immigrants to the USA, some of whom will have known that inside the statue is inscribed a passage containing the words *'Give me your tired, your poor, your huddled masses yearning to breathe free'*. America has received huge numbers of immigrants over the years, and welcomed them from Europe, including Martha Grier's own parents.

Emily Dickinson died.

> Hope is the thing with feathers
> That perches in the soul,
> And sings the tune without the words,
> And never stops at all

The **Wilson Ornithological Society** was founded, named after Alexander Wilson.

Coca Cola first sold, in a Georgia pharmacy.

1887

Aldo Leopold was born Burlington, Iowa. He died in Wisconsin in 1948 and his most famous work, *A Sand County Almanac*, was published the next year. He wrote, 'There are some who can live without wild things, and some who cannot. These essays are the delights and dilemmas of one who cannot.'

Leopold promoted the concept of wilderness – 'Of what avail are forty freedoms without a blank spot on the map?' – and of a land ethic:

> This sounds simple: do we not already sing our love for and obligation to the land of the free and the home of the brave? Yes, but just what and whom do we love? Certainly not the soil, which we are sending helter-skelter down river. Certainly not the waters, which we assume have no function except to turn turbines, float barges, and carry off sewage. Certainly not the plants, of which we exterminate whole communities without batting an eye. Certainly not the animals, of which we have already extirpated many of the largest and most beautiful species. A land ethic of course cannot prevent the alteration, management, and use of these 'resources,' but it does affirm their right to continued existence, and, at least in spots, their continued existence in a natural state. In short, a land ethic changes the role of *Homo sapiens* from conqueror of the land-community to plain member and citizen of it. It implies respect for his fellow-members, and also respect for the community as such.

In 1935, Leopold helped found the Wilderness Society, whose current vision is 'a vibrant continent that supports the long-term health of our land, water, people and wildlife'.

The **Dawes Act** (Senator Henry L. Dawes of Massachusetts) authorised the US president to survey Native American tribal lands, allot land for individual Native Americans, purchase any surplus land and distribute that excess to non-Native Americans. Its aim was both to assimilate Native Americans into the land-owning culture of the successful European invaders, and also to grab some of that land for further settlement.

1888
Another quiet year.

1889
North Dakota, South Dakota, Montana and Washington
become the 39th, 40th, 41st and 42nd states.
The **Oklahoma Territory was opened up to settlers**. The
New York Times called it 'the biggest race ever run in the United
States'.

1890
The detailed results of the 1890 **census** were lost in a fire. US
population 62,979,766 (compared with 23,191,876 in 1850), of
which 248,253 were Native Americans (compared with
400,764 in 1850).

The findings of this census persuaded the US Bureau of
Census to announce that there was no longer a frontier. **Idaho
and Wyoming** became the 43rd and 44th states, and for the
first time it was possible to travel from the Atlantic to the
Pacific, or vice versa, 'from sea to shining sea', through
unbroken US states – up until now the western states of
California, Nevada, Oregon and Washington had been
separated from the eastern states by a line of territories running
down the Great Plains from Idaho and Wyoming through Utah,
Arizona, New Mexico and Oklahoma.

Until this time there had always been an edge to the USA.
There was the 'civilised' America of towns and settled land,
and beyond was wilderness – which represented the land of
danger but also of opportunity. It's tempting to think that the
rapid expansion of the USA and its closing of its 'manifest
destiny' of stretching from the Atlantic Ocean to the Pacific
Ocean was driven by a yearning for adventure, an indomitable
spirit and a need to explore, and surely that must be true to a
large extent, but historian Louis Hacker gave a very different
interpretation as long ago as 1924:

It was an agricultural society without skill or resources. It committed all those sins which characterize a wasteful and ignorant husbandry. Grass seed was not sown for hay and as a result the farm animals had to forage for themselves in the forests; the fields were not permitted to lie in pasturage; a single crop was planted in the soil until the land was exhausted; the manure was not returned to the fields; only a small part of the farm was brought under cultivation, the rest being permitted to stand in timber. Instruments of cultivation were rude and clumsy and only too few, many of them being made on the farm. It is plain why the American frontier settler was on the move continually. It was, not his fear of a too close contact with the comforts and restraints of a civilized society that stirred him into a ceaseless activity, nor merely the chance of selling out at a profit to the coming wave of settlers; it was his wasting land that drove him on. Hunger was the goad. The pioneer farmer's ignorance, his inadequate facilities for cultivation, his limited means of transport necessitated his frequent changes of scene. He could succeed only with virgin soil.

Wounded Knee massacre. The last sad act of the main American Indian Wars was a massacre of mostly unarmed Native American men, and many women and children, at Wounded Knee, South Dakota. The most charitable explanation for the massacre was that it was triggered by a misunderstanding involving a deaf Native American reluctant to hand over his expensive rifle to the cavalry. Several of the younger Native Americans had concealed firearms. A scuffle led to shots, which led to the troops firing on the Native Americans. In the confusion and frenzy perhaps as many as 300 Native Americans were killed. Fleeing Native Americans, some mothers with babes in arms, were chased miles before being killed. Thirty-one soldiers died on the day or later from their injuries – many of them had been wounded by the indiscriminate fire of their own comrades. Black Elk (1863–1950) was an eyewitness to the killing:

I did not know then how much was ended. When I look back now from this high hill of my old age, I can still see the butchered women and children lying heaped and scattered all along the crooked gulch as plain as when I saw them with eyes young. And I can see that something else died there in the bloody mud, and was buried in the blizzard. A people's dream died there. It was a beautiful dream ... the nation's hope is broken and scattered. There is no center any longer, and the sacred tree is dead.

Death of Sitting Bull.

Sequoia and Yosemite National Parks were established in the Sierra Nevada Mountains of California.

The approximate **low point of the US Beaver population**. Before the European invasion the North American population was in the order of 200–400 million; nowadays it stands at around 15 million and is making a recovery.

US economic production exceeded that of Britain for the first time.

US annual lumber production was a still-increasing 23.5 million board feet.

1891

Herman Melville, the author of *Moby-Dick*, died and San Francisco whaling was nearing its peak. Massachusetts had a much longer whaling tradition, with Boston, Provincetown and New Bedford the main ports.

1892

Quatercentenary of Columbus setting foot nowhere near the USA in the Bahamas.

The **Sierra Club was formed**. Now the oldest and largest grassroots environmental organisation in the USA. Founded in San Francisco, California, by John Muir. Its mission is:

To explore, enjoy, and protect the wild places of the earth;
To practice and promote the responsible use of the earth's

ecosystems and resources; To educate and enlist humanity to protect and restore the quality of the natural and human environment; and to use all lawful means to carry out these objectives.

1893

Brad's drink was invented. Its name was changed to **Pepsi Cola** in 1898.

1894

Hawaii became an independent republic after its monarchy was overthrown by resident American and European businessmen. It was annexed by the USA in 1898 as a territory and became a state in 1959 – after Alaska in the same year. Hawaii has been called the extinction capital of the world, with around 45 avian species being driven to extinction after the arrival of Polynesians in AD 800 and another 20 or more species after Captain James Cook arrived in the late eighteenth century. Habitat destruction, introduced Rabbits destroying the vegetation, introduced rats eating the eggs, young and flightless adults of many species, and introduced avian malaria seem to have seen off most of these species – 11 of them since the US annexation (and five since statehood).

1895

'Babe' Ruth born in Baltimore, Maryland.

1896

Utah became the 45th state. Settled by a band of Mormons fleeing persecution in Illinois, where their founder Joseph Smith was killed, the area of the Great Salt Lake became an area for Mormons from across the world to congregate. Immigrants from Scandinavia and the UK crossed first the Atlantic and then the Great Plains to join the Mormon community. Polygyny, practised by some Mormons, became a major source of tension with the rest of the USA, and only when the Mormons officially suspended polygamy in 1890 was the way cleared for statehood.

1897

Martha Grier became a grandmother when Fannie gave birth to William R. Reilley (or Reley).

1898

The **Curtis Act** (Senator John Curtis of Kansas) extended the Dawes Act (1887) to the Indian Territory in what is now Oklahoma and removed around 90 million acres of land previously reserved for Native American use. Ironically, Senator Curtis was of largely Native American descent, and was the first of such descent to become a vice-president of the USA (1929–33), but his original Bill was greatly changed in its passage through committees of the House and Senate, leaving Curtis unhappy with the final Act that carries his name.

1899

Death of Simon Pokagon. Known as the Red Man's Longfellow, Pokagon was a writer and an advocate for Native American issues. In *The Red Man's Greeting* he wrote:

> On behalf of my people, the American Indians, I hereby declare to you, the pale-faced race that has usurped our lands and homes, that we have no spirit to celebrate with you the great Columbian Fair now being held in this Chicago city, the wonder of the world. No; sooner would we hold the high joy day over the graves of our departed than to celebrate our own funeral, the discovery of America. And while ... your hearts in admiration rejoice over the beauty and grandeur of this young republic and you say, 'behold the wonders wrought by our children in this foreign land,' do not forget that this success has been at the sacrifice of our homes and a once happy race.

Ernest Hemingway born in Oak Park, Illinois.

1900

Martha and George were still in Bridgeport, living in Main Street. George, now 62, was listed as a 'stationary engineer' (presumed to be a steam boiler operator, or power generation worker of some description). They owned their house; it was not rented or mortgaged.

The **Lacey Act** (John Lacey, a Republic Representative from Iowa) banned the sale of illegally hunted or harvested game, timber etc. This helped to enforce restrictions on hunting of species such as the Eskimo Curlew.

Raven and Snowshoe Hare extirpated from Ohio, as was the Passenger Pigeon with the death of Buttons at the hands of Press Clay Southworth.

US annual lumber production at an ever-increasing 35 million board feet.

First **hamburger** sold.

1901

Vice-president **Theodore Roosevelt becomes president** when President McKinley was shot in Buffalo, New York, after six months in office. Roosevelt was then elected as president in 1905 and served until 1909. He is often regarded as the greatest US president; he was certainly the foremost naturalist of them all.

Roosevelt was a conservationist with a utilitarian approach. He wanted nature conserved so that it did not go to waste, so that it was there for people to use in future. He summed up his position in a speech in 1910:

> Conservation means development as much as it does protection. I recognize the right and duty of this generation to develop and use the natural resources of our land but I do not recognize the right to waste them, or to rob, by wasteful use, the generations that come after us.

In 1912, Roosevelt made a speech in which he said 'There can be no greater issue than that of Conservation in this country' – but he went on to be in favour of reclaiming the southern swamps, developing Alaska and using the forests and grazing lands of the West.

John Muir's *Our National Parks* **was published**. Muir's view of nature conservation was different from his president's, although they saw eye to eye on many things. Muir was a preserver of nature for its own sake, and he saw wilderness as superior to civilisation: 'Wild is superior.'

1902

The **Rocky Mountain Locust was driven to extinction**. This species probably had a relatively small core breeding range in the valleys and meadows of the Rocky Mountains in states such as Colorado and Wyoming. When conditions were right it would sweep across the Great Plains devouring grass and crops. It was a huge economic pest throughout the mid-nineteenth century.

Jeffrey Lockwood's book *Locust*, published in 2004, describes the scientific detective story that led to better understanding of the reason for this species' demise. His journey took him to the Knifepoint Glacier, Wyoming, where he found locusts preserved in the ice that demonstrated that they had been superabundant in some years – enough for their insect bodies to form layers within the glaciers.

Lockwood tells of the calculations of Dr Albert Child when observing a swarm of locusts in June 1875 in Plattsmouth, Nebraska. Rather like Audubon and Wilson watching flocks of Passenger Pigeons before him, Child estimated the size of the swarm and its speed and found that it covered an area of 198,000 square miles (513,000 km^2) and contained around 3.5 trillion (3,500,000,000,000) individuals.

Despite being hugely abundant, its core breeding range was small. Conversion of these areas to alfalfa and other crops disturbed the soil which harboured the locusts' eggs. Changes to irrigation and drainage made flooding of the soil more likely, and trampling by domestic cattle may also have been a factor.

The Rocky Mountain Locust was driven to extinction not by human cunning in fighting it as a foe, but by the speed with which farmers changed the Rocky Mountain landscape. For thousands of years farmers have tried to wipe out locusts across the Middle East, Africa and Asia. In America they succeeded – by accident.

The demise of the Rocky Mountain Locust may hold the key to the lack of recovery of the Eskimo Curlew even after it received legal protection from excessive hunting. Eskimo Curlews migrated up through the Great Plains each spring, and perhaps they exploited the occasional bonanzas of food supplied by Rocky Mountain Locusts, and indeed other grasshopper species.

Ansel Adams, photographer of American National Parks and wildernesses, was born in San Francisco. He was to say:

> We all know the tragedy of the dustbowls, the cruel unforgivable erosions of the soil, the depletion of fish or game, and the shrinking of the noble forests. And we know that such catastrophes shrivel the spirit of the people … The wilderness is pushed back, man is everywhere. Solitude, so vital to the individual man, is almost nowhere.

The **Newlands Reclamation Act** funded irrigation of dry lands in 16 western states and territories.

1903

President **Roosevelt visited Yosemite with John Muir**, who persuaded him to take back management of the park to the federal government and away from the state of California. A proposal was first floated to dam the Hetch Hetchy Valley in the northern part of Yosemite National Park.
Wright brothers' first flight.

1904

Chestnut blight arrived in the USA on imported Asian trees. It spread at a rate of 80 kilometres per year and within a few decades practically all American Chestnut trees, around three billion of them, had been killed. The American Chestnut Foundation states that:

> The American chestnut was once one of the most important trees in the Eastern forest. In the heart of its range a count of trees would have turned up one chestnut for every four oaks, birches, maples and other hardwoods. Chestnut was so abundant on the dry ridge tops of the central Appalachians, their canopies were filled with creamy-white flowers in early summer; the mountains appeared snow-capped.

In virgin chestnut forests the trees averaged 1.5 metres in diameter and 30 metres in height with the first 15 metres or so branch-free, which made excellent timber.

White-tailed Deer and Wild Turkey extirpated from Ohio.

1905

The **National Audubon Society** was incorporated and named in honour of John James Audubon. Its mission is 'to conserve and restore natural ecosystems, focusing on birds, other wildlife, and their habitats for the benefit of humanity and the earth's biological diversity.' Its 500 local chapters engage its members and the public in lobbying, citizen science projects and educational work.

The **National Bison Society** was formed to lobby Congress to establish public Bison herds.

Texas alone held 800 million **Black-tailed Prairie Dogs**, and at around the same time the US population was estimated at around five billion individuals. This colonial, burrow-nesting rodent was abundant throughout the mixed-grass and short-grass prairie, although around this time large-scale poisoning programmes started to reduce its numbers. Nowadays only about 2% of its former range area is occupied and the total population is thought to be around 24 million individuals (a 99.5% decline). Prairie dogs have been reduced in numbers deliberately because they can harbour bubonic plague and because they are seen (perhaps correctly, but perhaps wrongly) as being competitors with cattle for grazing. Their burrows, droppings and grazing create areas of rich vegetation favoured by Bison, Pronghorn and cattle. Prairie dogs are the prey of a wide variety of predators from Golden Eagles to Black-footed Ferrets. The Black-tailed is the commonest of four US prairie dog species.

1906

The **San Francisco earthquake** and the fire that followed it killed 3,000 people and left about two-thirds of San Francisco's 400,000 inhabitants homeless. The financial cost of the disaster was around $10 billion at today's prices. The inadequacy of water resources to fight the fires increased pressure for the damming and flooding of the Hetch Hetchy Valley in Yosemite National Park. John Muir and the Sierra Club opposed this idea, with Muir saying 'Dam Hetch Hetchy! As well dam for water-tanks the people's cathedrals and churches, for no holier temple has ever been consecrated by the heart of man.' But in 1913 Woodrow Wilson signed the Raker Act, which authorised dam construction.

1907

Oklahoma became the 46th state and Indian Territory ceased to exist.

Marion Robert Morrison was born in Winterset, Iowa. The era of the cowboy had ended and the era of the cowboy film, in which Morrison would star as **John Wayne**, lay ahead.

1908

Butch Cassidy and the Sundance Kid died in a shoot-out in Bolivia.

The **Model T Ford** started rolling off the assembly lines in Detroit, Michigan, marking the beginning of the advent of affordable car transport for ordinary working people.

Charles H. Hudson, a farmer in Needham Heights, Massachusetts, wrote to his Congressional representative, **John Wingate Weeks**, imploring him to sponsor 'a national law put on all kinds of birds in every State in the country, as the gunners are shooting our birds that Nature put here.'

1909

Birth of **Wallace Stegner** in Lake Mills, Iowa. He coined the phrase 'the best idea we ever had' to describe the National Park concept, saying the parks were 'absolutely American, absolutely democratic, they reflect us at our best rather than our worst.'

1910

Martha Grier was listed as a widow, aged 72, on the 1910 census. She was still living at the house on Main Street, Bridgeport, where she remained until her death.

USA population 93 million, UK population 45 million (1911).

Extinction of the Carolina Parakeet in the wild. The only native North American parrot, the Carolina Parakeet was a green bird with a yellow head which mainly lived in riverine forests and spread out into the far west at times.

The Carolina Parakeet made two mistakes which contributed greatly to its extinction. The first was that it was a pest of agriculture, so farmers didn't like it. In some ways it was a help to the farmer because it ate some weed species, but it also fed on crops and thus it was often shot for its trouble. And that's where the second factor came in – when some of a flock of Carolina Parakeets were shot the rest of the flock would disappear for a short while but then came back to see what had happened to their flock-mates, leading to more deaths, and after a short while, and another return of the survivors, even more fatalities. They were easy to shoot.

The last remaining Carolina Parakeet, a male called Incas, died in a zoo on 21 February 1918. In fact, he died in Cincinnati Zoo, in the very same cage that had housed Martha and in which she had died a few years earlier. If you are an endangered species and you are ever offered a spot in that cage just remember that the place has 'form'.

The **death of Mark Twain** could no longer be said to be an exaggeration.

1911

The **Weeks Act** (Representative John W. Weeks of Massachusetts) built on the Forest Reserve Act of 1891 which had allowed retention of forests in state ownership. The Weeks Act allowed additions to the existing forest stock and was aimed at conserving forests of the eastern side of the USA. Importantly, funding was also made available for such purchases. By the end of 1980 more than 22 million acres (9 million ha) of land had been added, through purchase, to the national forest system in the eastern United States. It is said that 'no single law has been more important in the return of the forests to the eastern United States than the Weeks Act.'

The **Northern Fur Seal Treaty**, signed by Japan, Russia, Great Britain and the United States, ended the indiscriminate hunting of marine mammals, including Sea Otters and Northern Fur Seals.

1912

New Mexico and Arizona became states – the last of the lower 48.

1913

The **Weeks–McLean Act** (Representative John W. Weeks of Massachusetts and Senator George P. McClean of Connecticut) banned spring hunting of migratory birds and the importation of wild bird feathers for women's (or, I guess, men's) hats. This helped to protect species such as Eskimo Curlew and Wood Duck, onto which the guns had been trained with the decline and then the extinction of the Passenger Pigeon.

1914

Martha Grier, born in 1838, who had probably never been out
of Ohio all her life, died, aged 76 years, seven months and 24
days.

Martha, the last Passenger Pigeon in the world, who had never
been out of a cage all her life, died, aged somewhere between 17
and 29 years.

Martha Grier's three-score years and almost seventeen were spent
making a home and bringing up a family. She married her husband
George after he returned to their native state of Ohio after
campaigning through the South in the Civil War, under the overall
command of an Ohio General, Ulysses S. Grant, who was later to
become US president. We do not know whether Martha and
George knew each other before the Civil War, or how they met,
but for the rest of their lives they lived in eastern Ohio, in Belmont
County, very close to the Ohio River.

George was listed as an engineer in the 1860 census, but on his
return after the war he was, in 1870, a mail agent – which
presumably means that he was part of the growth of the postal
service which was knitting together the ever-expanding population
of the USA. The Pony Express was a decade in the past, the
telegraph service now crossed the continent and the mail service
was expanding. In 1880 the Grier family income came from
George's work as a clerk in a saw mill where he played a part in the
loss of forests in his state and in large parts of the USA as a whole,
which was the major cause of the decline of the Passenger Pigeon.

We get no glimpse of the Grier family in 1890 as the census
results are missing, but in 1900 a 62-year-old George is working as
a stationary engineer, presumably in a factory or some other form
of industrial development. Their son Charles is living a few doors
down the road from them with his wife Mary Anna (nee
Ochsenbeim) and their first child, one-year-old Hazel. Charles's
elder sisters Mary and Frances have moved away from the area but
we know that at least Frances is married and has started a family

too. By 1910 Martha is still living in the house on Main Street, alone now because George died in January 1904, and Charles, Mary Anna and their three children still live nearby.

When Martha died in 1914 Charles signed the death certificate and was able to supply the name of Martha's father as James Dean but not the name of his maternal grandmother or the places of birth of either grandparent. It seems that Martha didn't talk much about her parents to her children – or perhaps they didn't listen. We know from the 1880 census return that James Dean was born in Scotland and that Martha's mother was English.

In Martha Grier's life the US population increased six-fold, and immigration provided a large part of that increase. James Dean, perhaps accompanied by his English wife (or maybe they met on the voyage or in the USA), arrived as a Scottish immigrant in May 1830. George Grier's mother was from Ireland, but his father had been born in Pennsylvania. When Martha lived in Goshen in 1860, the census of that year showed that her neighbours were a mixed bunch. The heads of families and spouses were Ohio-born (44%), from US states east of Ohio (34%), from Britain and Ireland (7%) and from Germany and Switzerland (15%). At the same time, in Belmont County, further east, where George was living, the figures were 38% Ohio-born, 35% from eastern states, 13% from Britain and Ireland, 12% from Germany and less than 1% (each) from western states (Kentucky) and Canada. In the 1910 census, Martha's Belmont neighbours were 46% Ohio-born, 28% from eastern states, 3% from western states (Kentucky, Michigan, Illinois and Missouri), 7% from Britain and Ireland, 10% from Turkey and 6% from Germany, Austria, France, Belgium and Greece combined. Throughout their lives, the Griers lived near immigrants from Europe and the children of immigrants (as were they themselves), and with families who had spread westwards from more eastern states. These people of Ohio origin, or from other parts of the USA or from Europe, came to regard themselves and their neighbours all as Americans.

As the US population grew, it spread westwards. When Martha Grier was born, the USA consisted of 26 states and covered a land area of 2.6 million square kilometres. The westernmost states (Illinois, Missouri, Arkansas and Louisiana) were all in the valley of the Mississippi River. Texas was still a part of Mexico, as was California, and Florida had not yet achieved statehood.

When Martha Grier died, the USA, through a process of purchase, negotiation, occupation and war, was a country of 48 united states (and the territories of Alaska and Hawaii) occupying its current boundaries. Its land area had increased to 9.8 million square kilometres.

Some of the territory was acquired through war with Mexico – notably California and, after it had spent some time as an independent republic, Texas – while other areas of land were purchased or negotiated. There were wars and skirmishes in the Pacific (with Fiji, Japan and China), but the main conflicts during Martha's life were internal ones – the Civil War and the Native American Wars.

The American Civil War was not about territory; it was about the way of life that should exist in the ever-expanding USA. The events of Bleeding Kansas preceding the war showed that it would be impossible for the USA to grow as, in Lincoln's words, 'a house divided'. The Civil War, during which Sergeant George Grier travelled through the South, resolved the American model of land tenure and agriculture. After the Civil War, although the South experienced decades of economic adjustment and hardship, the nation took shape, with state after state being delineated, populated and admitted to the Union.

When Martha was born there was still a frontier where civilisation abutted the wilderness. When she was in her early fifties the constant movement west of people occupying the land closed the frontier. But it was not an empty land that they entered. It was occupied by Native Americans with a very different style of life from the European invaders. Those Native Americans were pushed aside, and their final defeat was at the hands of the settler and the politician as much as at the hands holding a Colt revolver or a Winchester rifle.

The 'American Indian Wars' occurred over the period 1775 to

1920 according to the US Bureau of the Census, and have been more than 40 in number. They cost the lives of about 19,000 white men, women and children, including those killed in individual combats, and the lives of about 30,000 Native Americans. In contrast, the Civil War cost around one million American lives, both soldiers and civilians.

In terms of headcount the American Indian Wars were pretty small beer – I don't want to appear heartless, but they were, even though they are the stuff of some excellent Western films. By the middle of the seventeenth century the Native American population had declined to around half a million in the USA (although estimates vary an awful lot), because the European invaders brought with them diseases that wiped out more than 90% of the native population in the first couple of centuries of European invasion. Smallpox and measles, new diseases to the New World, killed far more Native Americans than did the Seventh Cavalry – in fact they killed far more Native Americans than were available for the Seventh Cavalry to kill in the nineteenth century.

Native Americans were treated not as fellow members of the human race by the European invaders, but as a form of wild animal inconvenient to the plans of the Europeans. They were occupying the land that the Europeans wanted, and which was essential to the realisation of Thomas Jefferson's vision of the USA being a country of land-owning farmers. The year of Martha Grier's birth saw the Cherokee Trail of Tears, and by the time she died the Dawes Act had caused even those lands to which the Native Americans had been moved to be divided up and reduced in size. There no longer was a place on the map labelled Indian Territory, and any Native Americans who were not already dead were living in poverty or were being assimilated into the European population.

Land was allotted by the government and handed out to freed slaves and veterans of the Civil War – the frontier was pushed back by the advance of thousands of land-owning farmers. They had been preceded by explorers, travellers, adventurers, hunters and trappers, but it was the farmers whose spread across the continent truly marked the march of civilisation and secured the closing of the frontier. This was Progress, at least that was how it was seen at the time – and for the most part it is how it is seen now.

Martha Grier's life was almost certainly a better one than that of her mother, and her two daughters had easier lives still. Communications improved as telegraph and postal services were established, the railroad spread across the continent and production lines started to churn out cheap and affordable motor cars.

Martha Grier never had a vote, though her life spanned the terms of office of 20 US presidents, from Martin van Buren to Woodrow Wilson – but her daughters did, after the 19th Amendment was passed in 1920. George was able to vote in nine presidential elections, and we can speculate that his first such vote helped his fellow Ohioan, and former commanding officer, Ulysses S. Grant, into the presidency in 1868, and that his second helped keep Grant there in 1872.

If you measure the USA in Martha Grier's life by economic and social measures then it was a period of enormous success. The population increased, economic prosperity increased, the dominion of the rule of law increased, life expectancy increased, the ease of travel increased, and (unless you were a Native American) your ability to partake of life and liberty and to pursue happiness increased. This was Progress.

But, as we saw in the previous chapter, it was also Progress that drove the Passenger Pigeon to extinction.

Surely the child Martha Dean saw flocks of Passenger Pigeons snaking through the sky above her Ohio home. In some years they may even have nested in the nearby woods of Ohio, Pennsylvania or Virginia, and she may have seen feeding flocks regularly through the early summer. Was there sometimes a winter roost of millions of birds near her home? After all, that part of Ohio was famed for its Passenger Pigeon roosts. And did her father James go into the swampy woods and kill Passenger Pigeons in the ways described by Audubon?

Did Martha Dean eat pigeon pie? It seems likely that she must have done at some stage – perhaps it was a regular meal in the Dean household. Did father James get out his gun and bring down birds from a flock winging its way across the Ohio skies? As a wife and mother did Martha Grier cook pigeon pies for her family, and did they compliment her on her cooking?

Ohio, as we have already seen, played a notable part in the Passenger Pigeon's last years. The last Passenger Pigeon reliably recorded in the wild was killed there, about 200 miles from where Martha Grier lived, by Press Clay Southworth in 1900, when Martha was 62, and she was probably enjoying being a grandmother rather than bothering about Passenger Pigeons. In any case, at that time people did not fully understand, nor fully believe, that extinction in the wild had occurred.

Through her last years Martha Grier may have read reports that the Passenger Pigeon had ceased to exist in the wild, and she may have read about the last few left in captivity, and eventually the last single Passenger Pigeon. I wonder – did she know about Martha the Passenger Pigeon, the bird that shared her name? Did she feel any sort of connection with that last remaining Passenger Pigeon sitting alone in her cage 200 miles away in Cincinnati? Was she aware, in the years after her own husband George had died, that another Martha had lost another George? If she was, then she surely must have felt a pang of empathy.

Through Martha Grier's life, the wildlife of her native Ohio was under assault – often a deliberate assault. The Elk, Grey Wolf, Mountain Lion, Lynx, Fisher and American Marten were all extirpated from her home state before she reached her teens. The Black Bear hung on until her daughters' teenage years, and the Snowshoe Hare, Raven and Porcupine until her first grandchild was born. Ohio forest cover reached a low point in Martha's mid-thirties but was regaining ground in her later years. Did she, I wonder ever, visit Dysart Woods where I hugged an ancient tree? It was only a 20-minute drive for me but would have been a day's journey for her, I guess. And anyway, what would Martha Grier have wanted with an old tree except that it be turned into timber or firewood?

Martha Dean may have seen brightly coloured Carolina Parakeets, or heard them squawking, around her Coshocton home – they would surely have been noticeable – but they were gone from Ohio before she was married and gone from the USA as a whole before she died. More distantly, but in other parts of the ever-growing USA, there were other extinctions and near-extinctions – were they events that she read about or heard about? Did she pay any attention to them?

The ecological destruction that occurred in the USA in Martha Grier's lifetime was immense in scale and exceptionally rapid in execution. No other continent has experienced the extinction (or near-extinction) in the wild of so many widespread and formerly abundant species in such a short period of time. The years of Martha Grier's life encompassed the extinctions of Great Auk, Labrador Duck, Rocky Mountain Locust and Passenger Pigeon and the near extinction of the Bison, Pronghorn, Beaver, Sea Otter, Black-footed Ferret and Eskimo Curlew. And others were set on the road to extinction during her lifetime: Ivory-billed Woodpecker, Heath Hen, Xerces Blue Butterfly and Bachman's Warbler.

Different species, with different ecologies, but none was safe from the huge ecological changes that swept across the continent in a matter of decades. Forests were cut down and grasslands were fenced and ploughed. Species were harvested as if there was no limit to the numbers that could be taken. The mind-set was that nature would provide, however much we asked of her, and this was established by the size of the country and its original richness in nature. Europeans coming to the New World were amazed by the abundance they found there. They had no reason to believe that Passenger Pigeons could ever be shot to extinction. Why should they? At the beginning of the century which saw the Passenger Pigeon's extinction Audubon and Wilson were reporting single flocks that numbered billions. The Bison were abundant in a way that no Europeans had seen grazing animals in their home continent. The European invaders took what they wanted and counted their blessings.

And the ecological destruction was very rapid. Trillions of Rocky Mountain Locusts, billions of Passenger Pigeons and millions of Bison existed when Martha Grier was born, yet only 50 years later all three species approached extinction – and only in the case of the Bison was extinction averted. The pace of the final push across the continent after the Civil War was headlong. Nothing like it had happened in human history, and in particular nothing like it had happened in modern times when the men closing the frontier had modern guns, modern means of transport and the communications links to ship huge quantities of meat and skins back to markets, primarily in the east.

Just as the Passenger Pigeon's demise was made possible by modern communications networks, which spread the news of nesting colonies, and the expanding railroads, which provided the means to ship pigeon meat rapidly to growing markets, so was the slaughter of the Bison, Eskimo Curlew and Pronghorn. The scale of ecological destruction, and the speed with which it happened, were only possible because these were 'modern' times with ever-improving technologies of communication of information and transport of people and produce. This was Progress.

In fact, this was the 'tragedy of the commons' writ large across a continent. No-one owned the resources but everyone wanted them and so a market existed. Even if every man who shot a Passenger Pigeon or Bison knew that he was hastening the species' possible extinction (and most would not have believed it) there would still have been a huge incentive to keep firing away because the financial reward was immediate. And the same was true of forests and grassland too – they seemed limitless, and even if they weren't there was money to be made today, for sure, and if not by you then someone else would be making it.

It was a unique combination of technological ability, a growing market and a lack of legal restrictions which allowed this scale of destruction. And the political imperative was to push the frontier back and settle the land. Plenty of blind eyes were turned to how it was done, and the maintenance of natural riches and natural beauty was not high in people's minds.

Martha Grier was involved in the extinction of the Passenger Pigeon – she consumed resources which may have included pigeon pies, but in any case she was part of the population needing food, fuel and goods and creating the economic incentives for habitat destruction and species extinction. The Martha Griers of the late nineteenth century killed the Passenger Pigeon just as much as the men with guns in Petoskey. It's easier to blame the men in Petoskey, because they directly killed hundreds or thousands of birds, but at the other end of the supply chain were thousands of Martha Griers; people like you and me.

Towards the end of Martha Grier's life there was a growing environmental movement in the USA. The American Ornithologists' Union, the Sierra Club and the National Audubon Society were all formed. America's 'greatest idea' of National Parks came into existence and the federal government established laws to protect species and the habitats on which they, and we, ultimately depend. The birth of the American conservation movement was partly a reaction to the dramatic losses witnessed by the founders and partly a result of the increased living standards that allowed people to examine what was important in their lives and their leisure. And the USA produced great ecological thinkers and advocates at this time – John Muir, Theodore Roosevelt and Ansel Adams are outstanding examples.

The account of the USA between 1838 and 1914 which you (may) have just read is different from the one you will find in most history books – it has the ecology left in! Few books of American history deal with other than the economic, social and political aspects of life – what we might call the human froth on top of the real biological world in which we live. Some history books will mention the fate of the Bison, because it was wrapped up in the fate of some of the Native Americans, but you will struggle to find a mention of the Passenger Pigeon's extinction in history books – despite its global significance.

Putting the ecology back into history gives it a different slant.

Instead of the story of Martha Grier's life being one of continuous progress, it is a story of steady social and economic progress but at the expense of some ecological loss. Her life saw both slavery and the Passenger Pigeon abolished, and I wonder how much ecological loss could we cause and still be prepared to call it progress on the basis of the social and economic gains?

I'm struck by how recent it all was. My four grandparents were all born before 1900 (in 1889, 1893 and 1894), before the Passenger Pigeon was driven to extinction in the wild, and they were all in their twenties when Martha Grier and the last Passenger Pigeon died. My grandparents, had they been American, could easily have spoken to people who saw Martha in her cage in Cincinnati Zoo, or even have seen her themselves. They might even have spoken to people who could have given them first-hand accounts of the flocks of Passenger Pigeons, the herds of Bison and the loss of American forests and grasslands. Whereas they couldn't do the same back here in the UK or Europe because our destruction of our natural environment happened much longer ago and much more slowly and does not live in people's memories. The European destruction of wildlife lives in the history books, and much of it is fairly obscurely recorded there. We Europeans had already chased the wildness out of western Europe long before we moved in our millions and did the same in the USA. The ecological destruction which occurred in the USA through Martha Grier's life was massive, rapid and recent, and therefore it stands as a particularly clear and stark example of Progress.

The destruction of America's natural riches during Martha Grier's lifetime prompts us to ponder what Progress really is. That's a subject for the next chapter.

I found Martha Grier's grave in the Weeks Cemetery above Bridgeport, Ohio, where she had lived.

The Municipal Building in Bridgeport is on a rather grotty Main Street somewhere along which Martha had had her house. The Bait and Tackle shop looked the right age to have been standing when Martha walked these streets, but the Sylvania Color TV and Stereo store certainly wasn't. In fact the top floor of the building isn't standing now as it burned out, apparently some time ago. The Belmont Eagles, Aerie 995, has bingo on Mondays and Wednesdays at 6:20 pm. And just in case a visit to the Laws of Attraction salon and spa, which offers you a 'beautiful experience ', results in you having to fight off your admirers, never fear – it is next to the USA Martial Arts Center, which offers training in taekwondo, karate, aikido and self-defence.

I parked and entered the Municipal Building to seek a location for Martha's grave. At reception, a man was paying his water bill, or at least some of it. When he had finished I asked the receptionist where I could enquire and she, with a big smile, directed me to the mayor's office where a less smiley woman said she knew where to look. As she looked out the ledger I read the extract of a speech that was never given under a photograph, hanging on the wall, of the handsome young President John F. Kennedy.

The ledger said there was only one Grier grave, and that was an R. D. Grier who died long before Martha. I thanked the woman and headed out of town, up the Cadiz Road which winds steeply through woodland. On the sharpest left-hand bend there is a right turn signposted 'Shooters – Firearms and Accessories. 500 guns in stock.'

Weeks Cemetery on Sunset Heights is a well-maintained graveyard. As soon as I arrived I noticed a large pink column bearing the name Grier. I parked in the shade of a tree and walked across the newly cut grass to investigate. This was Ross D. Grier's memorial, and there were other names too – an Eleanor who had died in 1874 and a Mary M. Cooper who had died in 1890.

I looked more closely at the inscriptions, and I realised that Ross Grier was the son of George and Martha, and that he had been born on 9 March 1872 and died on 22 March 1874. So George and Martha had had another son, who did not show on the census records because his short life filled a few years in between the 1870 and 1880 censuses.

In the grass where I stood there were two headstones which said simply Father and Mother – these must be George's and Martha's last resting places.

If Martha could rise from her grave and stand next to me she would see that her plot was the prime spot in this cemetery – on the crest of the ridge between two attractive trees and with a view towards the Ohio River near where she had spent most of her life. The view cannot have changed much – it was still mostly trees – but as I stood there a Starling flew past, and there would be no flock of Passenger Pigeons, however long I waited.

I often return to Martha Grier's graveside in my thoughts – and it's always a sunny May day as it was when I visited. I wonder whether Martha Grier had a happy life. Was she deeply in love with George, and did he treat her well, or did he perhaps knock her around? Did she dance, did she sing? Was she house-proud? Was she a good mother? Did her children love and respect her? Was she, perhaps, a cranky old woman towards the end?

I often wonder whether Martha Grier ever saw a Passenger Pigeon, and whether she ever ate one. Did she know of their decline in numbers? Had she read about the Bison, and if so, what did she think of their slaughter? Did she care at all about the nature around her? And if she could have saved the Passenger Pigeon would she have wanted to do so? How much would she have cared? And how much do we care now?

CHAPTER SEVEN

The tolling bell?

No man is an island, entire of itself; every man is a piece of the continent, a part of the main. If a clod be washed away by the sea, Europe is the less, as well as if a promontory were, as well as if a manor of thy friend's or of thine own were: any man's death diminishes me, because I am involved in mankind, and therefore never send to know for whom the bell tolls; it tolls for thee.

John Donne

SHOULD WE REGRET THE EXTINCTION OF THE PASSENGER PIGEON?

On 22 November 1963, the idealism of America received a blow from which it has not yet fully recovered. President John F. Kennedy was shot while riding in a motorcade through Dealey Plaza in Dallas, Texas. The world would have been a different place if a flock of a hundred million wintering Passenger Pigeons had been flying over at the time, the air had been thick with their droppings, and the roof of the car had been up.

Almost exactly a century earlier, on 19 November 1863, Abraham Lincoln started his 271-word Gettysburg address with these words, 'Four score and seven years ago our fathers brought forth on this continent a new nation, conceived in liberty, and dedicated to the proposition that all men are created equal' – and finished with the wish and promise that 'government of the people, by the people, for the people, shall not perish from the earth.' Lincoln's warm-up act was Edward Everett, who spoke for two hours and whose opening remarks referred to the 'mighty Alleghenies dimly towering before us' – and he would have known that those mountains were one of the strongholds of the Passenger Pigeon, although neither he nor Lincoln would have guessed that it would be the Passenger Pigeon that would perish from the earth at the end of the century.

Lincoln and Kennedy were two presidents, both assassinated in office, who were positioned symmetrically on either side of the extinction of the Passenger Pigeon. One who lived through the Civil War and the other who lived in the Cold War. One who must certainly have seen Passenger Pigeons and the other who most certainly did not. One who seems like just another character in history to me and the other whose death I can remember even though I was a young child at the time. Go back another 50 years from Lincoln's time and we are in the presidency of Monroe, when Alexander Wilson had produced his *American Ornithology*; go forward 50 years from Kennedy and we are in the presidency of today, of Barack Obama.

Should we just accept that the Passenger Pigeon belonged to the age of Monroe and Lincoln, and that regretting its passing makes as much sense as wishing that Monroe or Lincoln were sitting in the Oval Office? Is the extinction of the Passenger Pigeon a sign of Progress that we should cherish rather than a loss that we should regret? Is the Passenger Pigeon like the stovepipe hat, perfectly acceptable and welcome in the nineteenth century but anachronistic in the twenty-first?

Our answers will depend on what sort of world we wish to live in. I'll tell you a little about what sort of world I would like to see, but first let's recognise that the Passenger Pigeon's extinction was regretted at the time.

Those who knew the Passenger Pigeon, who had seen it, shot it, eaten it, and watched it snake through the skies in astounding numbers, regretted its passing. At the time of its terminal decline some argued that measures should be put in place to protect it from overexploitation, and as these measures failed, rewards were offered for evidence of its continued existence. People wrote their memories of the Passenger Pigeon and they were almost always fond memories, and memories of its beauty and unparalleled abundance.

No-one sought the Passenger Pigeon's extinction and many

mourned its loss. Only a few voices were raised to speak ill of the dead and to express relief that this annoyance to farmers was gone from the land. We let the Passenger Pigeon slip through our fingers and for the most part regretted that we failed to hang on to it.

If the Passenger Pigeon was still with us – perhaps, to be realistic, in much smaller numbers than in its heyday – then it would provide an unmatchable wildlife spectacle that Americans could be proud of and tourists could enjoy. I would have sought out the Passenger Pigeon wherever it was nesting in May 2013, and I would have gone to wonder at it.

Maybe Petoskey and Wisconsin Dells could cash in on the Passenger Pigeon in a completely different way after all these years, but the unpredictability of the species' arrivals and departures would make it difficult to build local tourism-related businesses around the bird. This season they might be in Michigan for a few weeks – but next year perhaps it would be Canada or New York.

But human ingenuity would find a way around these practical problems. There would be a pigeon app for my iPhone and a 'Pigeon Watch' website, rather like 'Tornado Watch' but giving news of the presence of the biological storm. People would flock to see clouds of Passenger Pigeons at nesting sites, feeding sites and winter roosts. They would delight in seeing the males arriving mid-morning to relieve the females in the incubation period and then all the females returning a few hours later. The to-ing and fro-ing of the birds as they fed their young would be one of the most frenetic wildlife spectacles imaginable. When the adults departed (and what a sight that would be) then tours would take you through the colony to see the bloated squabs on the ground and to appreciate the density of nests scattered through the trees.

Your experiences of Passenger Pigeons would be mentioned in the same conversations as the Wildebeest migrations of the Serengeti and the penguin colonies of Antarctica. They would be part of the one-upmanship of travellers' tales at dinner parties. And yet these would be tales from the eastern USA where millions of people live.

Americans are keen birders – a study published in 2003 revealed that in the USA 46 million people watch birds – nearly one in five adults – and they spend $32 billion in retail sales annually, thereby contributing $85 billion in economic output and supporting 863,405 jobs. A few living Passenger Pigeons could only add to those totals.

I remember meeting Pam and Darryl at Magee Marsh, Ohio, in 'The Biggest Week' in 2011. Their Travelling Café sells burgers, hot-dogs, coffee, cakes and cold drinks to the hundreds of birders who visit this spot to get amazingly good views of spring warblers heading north. Pam, a short, slim, middle-aged woman with a ready smile, is front of house – taking orders, being nice to the customers and taking the money – whereas Darryl, who looks like he may have tasted his own cooking (and quite right too), fries away outside the van.

Pam told me they are based in Columbus, Ohio, and spend 10 days at Magee Marsh each year – and it's good business. They usually sell out of hot-dogs (200 or more) by noon. The rest of the year they do auctions, street fairs and art shows, but Pam said she liked the birders best – I bet she says that to all the guys. She and Darryl would be able to follow the birders and the Passenger Pigeons if they were still around, and they would be part of a small industry based on the bird.

We would be able to buy Passenger Pigeon mouse-mats, Passenger Pigeon-emblazoned baseball caps, Passenger Pigeon mugs and a whole range of Passenger Pigeon books to delight and enthral us. Local farmers would let out their fields for camping while the birds were in town, and local restaurants and diners would do a roaring trade.

Perhaps pillows stuffed with Passenger Pigeon feathers would be on sale – they were supposed to have miraculous life-prolonging properties.

Considering that the Passenger Pigeon has been extinct for nearly a century it is remarkable that there are a few songs about it, and we would have our pick of them to play on our iPods as we drove to see the birds. The Handsome Family's song 'Passenger Pigeons' is one of their least popular tracks on Spotify, even though it expresses disbelief as to how billions of birds can disappear; I prefer the acoustic guitar solo, also called 'Passenger Pigeons', by Matthew Gennaro on his album *The Invisible Pyramid: Elegy Box*. Another 'Passenger Pigeons' is by Resonator, a two-minute instrumental on their album *Lost Language*. There is also a group called Passenger Pigeons, whose *So Far … So OK* album has tracks such as 'Two Struggling Sparrows' and 'Slow as Snails'.   And Martha has her own song, 'Martha (Last of the Passenger Pigeons)', by the late John Herald, which laments her passing. But top of my list is the Barbara Dickson song 'Passenger Pigeon', which is the only place, I believe, where you will find the species' scientific name, *Ectopistes migratorius*, actually sung.

In theory, we could harvest Passenger Pigeons for food, sustainably, maybe under licence. A meal of 'organic, free-range, wild and local' Passenger Pigeon might never replace the traditional American Thanksgiving dinner of Wild Turkey but on Buttons Day (24 March) and Martha Day (1 September) maybe some families would sit down to a meal of pigeon pie.

However, we would be fooling ourselves if we believed that the Passenger Pigeon would be welcomed back gladly by all. These days its ability to make even a local difference to cereal crops would be small, as Passenger Pigeons took grain sown onto the surface of the soil – the seed drill will have sorted that out. I would be amazed, though, if there weren't some complaints from local farmers if the pigeons were still around. We would hear that they were eating crops and scaring livestock, and no doubt claims for compensation would follow.

Foresters and woodland owners might have more to worry about. The prospect of your wood being wrecked by a visit from millions of Passenger Pigeons would not be attractive. Would landowners call for a cull?

Audubon wrote of the dung falling like melting flakes of snow, and the atmosphere in Louisville being strongly impregnated with the odour of the birds. This might not be quite the experience sought by spectators, dressed in their finery, at the Kentucky Derby on Churchill Downs, Louisville, in the first week of May each year. Imagine all those fine hats and white suits speckled with pigeon droppings.

Can you imagine the uproar if scheduled flights into and out of Chicago O'Hare International Airport were disrupted for six hours by the passage of tens of millions of Passenger Pigeons?

There are two landscape-scale ecological consequences of the Passenger Pigeon's demise which we still live with. Both stem from the impact of the bird on the ecology of northeastern forests through its huge numbers and its dependence on tree mast.

The colonies and winter roosts of Passenger Pigeons were important agents of disturbance, and patches of forest were destroyed by the breaking off of tree limbs and the deposition of huge quantities of dung. The large, well-fertilised forest openings, with increased quantities of twigs and branches broken from the trees by the pigeons, and standing dead timber, increased the incidence of forest fires caused by lightning strikes, and the intensity of those fires. Red Oaks are less able to withstand fires than are White Oaks, because of their thinner bark, and so the forest disruption caused by the massive roosts and colonies favoured White Oaks in competition with Red Oaks, and the extinction of the Passenger Pigeon swung the advantage back towards the Red Oak.

Red Oak acorns were also a more favoured food for Passenger Pigeons than were those of White Oak, simply because they persisted into the spring whereas White Oak acorns sprout in autumn. Once Passenger Pigeons declined to extinction, far more Red Oak acorns survived to grow into young trees. The removal of billions of acorn-eating Passenger Pigeons was another factor that shifted the balance towards Red Oaks in competition with White Oaks, and the composition of American forests changed as a result, no doubt with impacts on many other species which we have yet to identify.

A further consequence of the Passenger Pigeon's extinction was that other species feeding on acorns and beech mast survived better and their population levels rose. Deer and rodents would have been the main beneficiaries. As deer and rodents are reservoirs of the tick-borne Lyme disease, its incidence is thought to have increased. Lyme disease has an economic cost to the US human population estimated at $1 billion per annum. Allocating a part of that cost to the extinction of the Passenger Pigeon would make its loss a rather expensive one.

This litany of consequences of the extinction of the Passenger Pigeon is in danger of turning into a credit and debit account for an extinct bird. Our feelings about extinction should be nothing so mechanical and cold.

There are two main and overlapping ways of looking at the 'value' of nature and whether we should conserve it: one is a utilitarian view (what can the Passenger Pigeon do for me?) and the other is a cultural and moral view (what should I do for the Passenger Pigeon?). And at around the time of the Passenger Pigeon's extinction these two views could be personified by President Theodore Roosevelt, whose conservation was of quite a utilitarian bent, and John Muir, who was more strongly in the moral camp. Their different starting points often led them to similar conclusions, for example in the National Park concept, but they were on opposing sides when it came to the damming of the Hetch Hetchy Valley in Yosemite National Park.

We need to have a foot in both camps, in my view – but I lean strongly one way. This is how I look at these issues.

I start from being on nature's side. What makes this planet so very special is that it is, as far as we know, the only place anywhere that holds life. It is certainly the only place where you can see a Blue Whale or a Giant Sequoia. I saw both, for the first time, in the summer of 2013 and my life was hugely enhanced by both experiences. Both are amazing examples of life on Earth. They are both big and majestic and they inspire awe – in me at least. Both are much rarer than they once were, although both have been saved, at present, and only just, from extinction. I am not going to tell you that the reason I'm glad we saved the Blue Whale is so that we can build up its numbers and start harpooning it again. Its value to me is not in blubber or whale meat. And I won't tell you that my sadness at the lack of Giant Sequoia is because we could really make a killing by cutting them down if they were all still standing. Certainly not!

I value such species because of their beauty and because they are part of the amazing variety of life on the planet. They are some of Earth's 'signature species' – the species that an extraterrestrial would talk about on his return home from a holiday on Earth. They aren't the most important species, they aren't the only important species, but they are some of the most amazing species and the most accessible species. They are species about which most children and most adults would say 'Wow!' if they saw them, and the encounters would remain in their memories as treasured experiences. They are species that open our eyes to nature so that we can find the other less obviously dramatic species fascinating too. They are species that make us glad to be alive, and to be part of life on Earth.

And the Passenger Pigeon was a signature species, too. It had the 'Wow!' factor in spades! Audubon and Wilson didn't write of the Henslow's Sparrow in the same terms as they did of the Passenger Pigeon. The Passenger Pigeon was a biological phenomenon, a biological storm, and it amazed and impressed us in its time. How could anyone not wish that it were still writing its name in the skies of the USA?

If you ask me to put a $-sign on the Passenger Pigeon I'm not going to play that game. Yes, we could try to estimate the tourism value of the Passenger Pigeon and assess whether it would be more or less than that of Disneyworld. And we could try to estimate the increased cost of medical insurance and medical bills because of Lyme disease. And we might have to take off something for delays to flights because of Passenger Pigeon disruption – but this would all be nonsense, utter nonsense! The Passenger Pigeon was a signature species and we are diminished by its loss, but we will diminish ourselves even more if we regard its loss in solely economic, bean-counting terms. Let us lift up our eyes to the skies and imagine an enormous flock of Passenger Pigeons and proclaim that the world would be a better place if there were still a real prospect of seeing them. Yes, I regret the loss of the Passenger

Pigeon – I regret it deeply and passionately because we lost a significant element of the living planet's natural beauty when we drove it out of our lives for ever. I feel robbed and I feel saddened by what we did. Did you expect me to say anything else?

WOULD WE BRING BACK THE PASSENGER PIGEON TODAY?

Dan Novak and Beth Shapiro from the University of California at Santa Cruz are sequencing the genome of the Passenger Pigeon from fragments of DNA from museum specimens, and they plan to use Band-tailed Pigeons to fill in any missing bits and to be the 'foster parents' of almost-complete Passenger Pigeons. There are huge technical difficulties in accomplishing this genetic task, but if they pull it off there is the prospect that there might be another cage, sometime, with a group of 'reborn Marthas' in it. Would Cincinnati Zoo host these new birds?

I'm not sure how I would feel looking at a cage of reconstituted Passenger Pigeons. I think there might be some of the same sadness that I would have felt in seeing the original Martha caged and alone when she should have been just one of a flock of millions.

It's difficult to conceive that the Passenger Pigeon could be brought back to our skies in millions, and if that isn't the potential endpoint then I think I'd rather we didn't bother at all, and spent the money on a project with a more attractive finale.

If the species were still with us it would be very different, for me at least, from us putting it back into the environment after all this time. If the Passenger Pigeon had survived it would show that we had found a means of living with it, coming to some sort of an accommodation with it. It would mean, possibly, that when our flights were delayed by pigeons we would be accustomed to it – not pleased by it, but we would accept it as one of those things like tornadoes or snow that we just have to work around (maybe we would even be offered insurance against delay by Passenger Pigeons when we booked our tickets online).

If the Passenger Pigeon was still with us then it would be a different world. Perhaps it would mean that we had planted new forests as reserves for Passenger Pigeons over the last century. Maybe we would have set aside some areas of land where we fed Passenger Pigeons on grain or corn, as we do cranes and geese these days (to lure them away from commercial crops, but also to help conserve them and provide a spectacle for the viewing public). Maybe it would mean that we had licensed pigeon hunting and were ploughing back the proceeds from licence fees to compensate landowners for pigeon damage. Or maybe it would mean that we had decided that we loved them and couldn't do without the Passenger Pigeon in our lives.

A GLIMPSE OF REMAINING ABUNDANCE

On my last evening in the USA in the summer of 2013, I was standing on the Congress Avenue Bridge in the middle of Austin, Texas, with the Colorado River below me. This was in the middle of a modern American city. The headquarters of the *Austin American-Statesman* newspaper were to the south of the bridge near the Hyatt Hotel, and on the north side of the bridge were tall buildings including the Austonian, 360 Condominiums and the Frost Bank Tower.

I wasn't alone. The bridge was packed with people – I would guess around 500 of us – and there were another 300 or so on the banks of the river and yet another 100 in boats on the river. Street vendors sold us ice-creams and bottled water. There were all sorts of people there. Four young women were texting their boyfriends about where to meet later, and they looked set for a good night out. Grandparents, parents and children waited together for a wildlife spectacle to start. A young woman in cut-off shorts, with a deep tan, a scooped neckline, red lipstick, white teeth and light green fingernails with matching toenails said to the two drippy-looking blokes with her that 'things should start right now' at 8.38 pm. She didn't look like a wildlife expert to me – but I was wrong, and within 90 seconds thousands of Mexican Free-tailed Bats were emerging from under the bridge on which we stood. A few metres below me there were suddenly thousands of bats swirling around.

A plume of bats headed out from the bridge for a couple of minutes – an unending stream. Wow! That was great! But that was just the beginning.

After a short break the stream started again, with bats pouring along the south bank of the river. Then another stream started from the middle of the bridge and another from the northern side. Three plumes of bats heading off to feed. We are told that about 1.5 million bats use this bridge as their nursery colony. I'm not arguing with that – there were lots.

This was a natural wonder. It was free, and it was in the middle of one of the USA's big cities, and it happens every night through the summer. It was the most memorable wildlife spectacle that I have ever witnessed. I would have paid to see it, and I would pay now so that others could see it if I were sure that my contribution would help this natural wonder continue into the distant future.

The bats were so close to me under the bridge. I could smell their musty odour. Looking down I could see bats swirling around and around in a tight circle, mostly under the bridge and out of sight but the rim of the circle extended into my gaze. It looked like part of an enormous dark machine spinning powerfully in the fading light as thousands upon thousands of bats circled around in a tight coordinated movement – each following the same circle, each holding its course and each moving at the same speed. Out of my sight there must have been baby bats huddling together as their mothers took to the air and performed these aerial manoeuvres, but I could see nothing of them. What I could see was a stream of bats pouring out from under the bridge. Looking down on them against the dark water of the river they were just discernible, and only as a darkening of the gloom, but as they moved over the tops of the riverside trees I could see their wings flapping and their bodies moving through the air. Hundreds passed each second, creating the impression of smoke.

I could see the bat-plume for maybe two kilometres until it disappeared behind trees and buildings and into the twilight gloom. Like smoke, the stream of bats was unbroken, stretching into the distance. And like smoke it seemed, in the distance, to blow in the wind and slightly alter its course, but it always remained an unbroken plume stretching from the bridge, above the riverside trees and into the distance. And out of sight, the front of this plume was still heading towards a distant feeding ground, getting further and further away as I watched, but still connected to the colony beneath my feet.

The impression of vast abundance will stay with me for ever, but so, also, will a little event which I suspect hardly anyone else noticed. As I watched the distant bats pass over the trees, a Peregrine Falcon flew in from the left and plucked a bat from the plume and flew back into the modern skyscrapers of Austin on the north bank of the river. It wasn't a dramatic capture, the Peregrine didn't dive at speed into the bat-plume, it simply flew alongside and casually, it seemed, picked a bat out of the crowd and flew back across the river – perhaps as it did every evening throughout the summer at about this time. I could see the captured bat's wings beating as it was held in the Peregrine's clasp, but the plume continued unaffected. And the presence of the falcon affected the passage of the plume not one bit. The bats took no avoiding action; they reacted not at all to the Peregrine's arrival or departure. This was death for one bat, dinner for one Peregrine, but not even a distraction for all the other bats. This was the dilution effect captured in an instant for me.

Maybe I saw three-quarters of a million bats in half an hour that evening. I really don't know, and I really don't care how many there were. The numbers were astounding and the sight of huge numbers of bats flying up the river with tall buildings as a background, cars passing us on the bridge, and people taking photographs with their mobile phones was an encouraging example of wildlife abundance in modern America. There were lots of people enjoying the spectacle and leaving enthused and happy. This was an uplifting experience and a helpful one in trying to imagine what Audubon and Wilson might have seen two centuries before, even though their flocks of abundance had been more than a thousand times bigger. But it helped me imagine what an enormous flock of Passenger Pigeons would have looked like as it stretched far into the distance, as did the stream of bats. It also reminded me that people enjoy the sight of wildlife abundance if it is available to them. And amidst the tall buildings of modern America it made me hope that wildlife abundance and Progress were not incompatible.

PROGRESS AND THE LOSS OF ABUNDANCE

I had come to the USA to seek understanding of lost abundance, and was leaving the USA with thoughts of current abundance. I regret the extinction of the Passenger Pigeon but we probably can't bring it back. And in any case, nature conservation must live in the present – in the world of the Mexican Free-tailed Bat which hangs on, admittedly in diminished numbers, but still in abundance, even in the middle of modern-day Austin, Texas.

As I arrived at George Bush Intercontinental Airport I was thinking of the USA that I was leaving. As I pulled in to the bay to return my hire car a man scanned a sticker on the window of the car, looked at his machine, and as I lowered the window said, 'Hello Mr Avery. You've had a long trip. Welcome back to Houston, I hope everything was all right with the vehicle.' This was so American: efficient technology making life easy and a friendly welcome with good customer care.

I took my two bags out of the car in which I had spent much of the last five weeks. It was a car which told me how much fuel I was using and I had tried to drive to keep fuel consumption down throughout my trip. It had air conditioning which had been a boon in the deserts but which I had tried not to use too much. I had charged my British mobile phone and my American cell phone in the car and used them to phone ahead to book accommodation and to wish my son a happy birthday from a desert in Utah. The satellite navigation kit I had bought in Arkansas had guided me around 20 US states and back to my starting point in Houston, and satellite radio had allowed me to listen to the music of the 1970s and 80s as I travelled through the America of today.

The great great grandchildren of Martha Grier have comforts

and luxuries which she did not imagine – of that we can be sure. And they live in a society which is largely democratic and caring and tolerant. Overall, I firmly believe that I live in a better world than the one that Martha Grier inhabited. Our species has much to be proud of.

Proud, yes, but not smug. For my contention is that it would have been an even better world if we had made different choices over the period between Martha Grier's life and our own. And also, that the future should not be a simple extension of the past.

Let us first look backwards. The uncomfortable thing for any environmentalist about the changes that occurred through Martha Grier's lifetime was that, although life almost certainly became better for the average American, this improvement was accompanied by loss of environmental riches on a massive scale: global extinctions, loss of wildlife, deforestation and removal of grasslands. Let us go back to what we know about the Passenger Pigeon. We drove the Passenger Pigeon to the brink of extinction because we cut down its forests and ate millions of pigeons. But we could have cut down lots of forests and eaten millions of Passenger Pigeons without driving it extinct. In retrospect, which is easy, we could have harvested the pigeons sustainably and still be eating them now (if we really wanted to), and we could have harvested the forests differently and left some areas that would still hold pigeons (at least it's a possibility). In the 1850s we didn't have the knowledge to do any of that – we can't call it a mistake made by Martha's generation, even though I might wish they had behaved differently.

Here's another example from Martha's lifetime. Imagine that, again in the 1850s, the USA had created an enormous Prairie National Park stretching from Kansas to Montana – let's say 200 hundred miles wide and 1,000 miles long (about 500,000 km^2). If that had happened, the USA would be even prouder of the idea of National Parks than it is today. Native Americans would have adapted their ways of life to fit better into the modern-day USA and there would have been a different reality that we would now accept. That different reality would include many more Bison, Pronghorns, prairie dogs and Black-footed Ferrets and just maybe some Eskimo Curlew, too. It would be richer in wildlife – by miles. We would also be celebrating the huge amounts of carbon that were stored in the prairie soils and had not been released into the atmosphere. And do you know what – no-one would be complaining now. No, shifting baselines work both ways – we don't miss what we haven't known and we don't resent what is familiar. If we had saved more of the Great Plains then we would regard that as being normal now.

Can we really imagine that taking out well under 10% of the US land mass and keeping it natural would have had an impact anything remotely like a 10% reduction of the US economy? Silicon Valley would still be there, the car and aeroplane manufacturing would still have happened, food production would have been reduced by far less than 10% and the USA would still be more or less the superpower it is now. If we had saved more of the environment then we would not have been noticeably less wealthy and we might just have been a little happier. I think that it was along similar lines that Aldo Leopold was thinking when he wrote:

> Our grandfathers were less well-housed, well-fed, well-clothed than we are. The strivings by which they bettered their lot are also those which deprived us of pigeons. Perhaps we now grieve because we are not sure, in our hearts, that we have gained by the exchange. The gadgets of industry bring us more comforts than the pigeons did, but do they add as much to the glory of the spring?

If we come forward from the year of Martha Grier's death to the present day then we see more social and economic progress. But the environmental scorecard must record growing realisation that American lifestyles are depleting natural resources and increasing pollution (including greenhouse gas pollution) to an unsustainable extent. Because of the ever-growing population, multiplied by the ever-growing per capita economic wealth of the USA (more people using much more 'stuff'), the present position is worse than when Martha died rather than better. Desert cities such as Tucson, Arizona, may run out of water in a few decades' time if present trends continue. California's fruit production in the San Joaquin Valley, once an enormous natural wetland, depends on irrigation too – and that water can't be used on the land, be drunk in fizzy drinks and fill swimming pools all at the same time. Will a future president decide to flood the Yosemite Valley, as a previous one decided to flood nearby Hetch Hetchy?

America is a massive polluter of the global atmosphere and leads the world in its cumulative per capita historical impact on the atmosphere. Much of that came from the last century, but the impacts will be felt in future, and carbon emissions continue to be, per capita, some of the highest in the world. The impacts of climate change will be found to affect every aspect of American life – and the emissions of America will affect not just America but also the rest of the world. There is one global atmosphere, and when we harm it we harm everyone.

Climate change is just one example of where America's actions have global consequences. The last century has seen a shift in the reach of the ecological damage – it is no longer felt just at home in the USA but abroad as well. Just as Martha Grier never chopped down a tree herself, but was part of the market economy that caused that tree's destruction, so is the modern-day USA a massive communal Martha creating a market which causes rainforest destruction in far-away corners of the globe. The scale of this impact is huge.

I've always liked the passage from John Donne that heads this chapter. Many years ago, when I was an undergraduate, it was stuck up on the wall of my room and it struck a chord with me then (and also made me feel rather sophisticated – which I wasn't really). It still strikes a chord with me (and I'm still not very sophisticated).

I am involved in mankind. I love nature but I care about the lot of my fellow man and woman too. Any man's oppression or poverty does diminish me, and all of us, morally, even if sometimes it enriches us materially. I believe that much of our future energy should be devoted to making the world a fairer place rather than a richer one. We have so much, but many people in the world don't get anything like the full benefits of our economic wealth. To my mind this applies to the people in my street, in my country and on my planet.

I do, also, feel that any species' death, like that of the Passenger Pigeon, diminishes us. I believe it reduces our moral authority because so many species extinctions are careless and unnecessary. If we continue laying waste to the diversity of life and the natural beauty on the planet then we are like vandals going through an art gallery with spray-paint and Stanley knives. When we come out the other side, or when we grow up, we look back and wonder why we didn't behave with more restraint, and we realise that the excitement and enjoyment that we had was rather twisted and in no way was worth the damage we did. Nobody gains from our vandalism and we are ourselves diminished.

Another analogy I use is to compare species to pieces of music. As we lose nature from the world around us it is like removing pieces of music from our lives. When a species declines then the volume of that piece is turned down and the sound is distorted. When extinction happens the music is silenced forever. I want nature in my life like I want music in my life. I don't expect to come up with an economic justification for the presence of music, and nor do I for nature. When we lost the Passenger Pigeon, a signature species, we lost a major symphony. I am tempted to say Beethoven's Seventh, but given the number of voices we lost with the Passenger Pigeon it might have to be the Ninth.

There is a sense in which any species extinction tolls the bell for our own demise. However, in our gloomy way, we conservationists tend to overstate this case. We say that the world will fall apart if we allow the extinction of species, but that clearly, so far, hasn't happened. The demise of the Passenger Pigeon did not toll the bell for the end of prosperity for the growing USA. And many other species have been lost without us feeling much pain, let alone feeling we have a terminal illness. The waitresses who served me breakfast in diners across the USA are not worse off economically because of the extinction of the Passenger Pigeon, so we might ask 'Why bother to save species from extinction?' The answer is partly because those waitresses would be no worse off economically if the

Passenger Pigeon were still with us. If it makes little difference to our own self-interest then let us choose the route of lesser environmental damage.

But this is where the past is a bad predictor of the future, I believe. We are now in the era of climate change, deforestation, soil degradation, water shortages and an unparalleled rate of extinction across the world. The bell started tolling many years ago – it is a warning bell rather than an expression of demise. The extinction of the Passenger Pigeon raised the volume of that bell a little a century ago, but now is the time for us to heed the warning.

We cannot know what President Kennedy might have done for the environment had he survived – maybe very little. But we do know that he was prepared to shoot for the Moon – in a famous speech he committed the USA to winning the space race (although it was lagging behind at the time) by landing a man on the Moon and bringing him safely back to Earth within a decade. And America achieved that goal, although Kennedy did not live to see it. If a current-day US president was making a speech that was shooting for the moon then I would hope that in view of the current environmental crisis it would have a more earthly focus, and I would hope that it might be a little like this:

President Obama stands, alone, by the statue of the Passenger Pigeon in Cincinnati Zoo, and looking straight into the lens of a camera, he says:

My fellow Americans. Today I pledge the United States of America to a new course in environmental protection and enhancement.

One hundred years ago today, almost to the very hour, America lost one of its number. In the small building behind me [he gestures] a bird died. She had been named Martha after Martha Washington, and she was a Passenger Pigeon. She would have looked to many of us very like the town pigeons that occupy our streets but she was the last of her line – the last Passenger Pigeon on Earth. When she died so did her species.

The Passenger Pigeon was a thoroughly American species – it lived only in the eastern parts of the United States where oak and beech forests predominated, and in the southern forests of our neighbour Canada. The Passenger Pigeon was the commonest bird on the planet, numbered in billions as we are now. So why is it no longer sharing our country with us?

[The president looks down at the bronze statue of the Passenger Pigeon next to him and puts his hand on its smooth head.]

Martha and the other billions of Passenger Pigeons were victims of the way we occupied, settled and developed our country. We cut down the forests in which Passenger Pigeons lived and we killed millions of them for food, and we did all this with such force and overwhelming power that we removed the Passenger Pigeon from America and from our planet in a matter of decades. We did not intend it, but the Passenger Pigeon was a casualty of progress. We caused the extinction of the commonest bird the Earth has ever seen in a few short decades of our history.

The Passenger Pigeon was not alone. Other American species were driven to extinction, and others such as the California Condor and the Bison, iconic species of our western deserts and plains, came perilously close to it. The exploitation of our seas, our forests, our rivers, and our grasslands was dramatic. America played its part in bringing the great whales close to extinction, we suffered the Dust Bowls of the 1930s because of poor soil conservation, and we lost most of the greatest trees on the planet, the Giant Sequoias of California. The Passenger Pigeon, and the last Passenger Pigeon Martha, stand as a symbol to all that destruction.

I am immensely proud of the economic richness that was built during the nineteenth century. Great Americans built the economic base of our country and improved the wealth of our people. We benefit from the riches they helped to create. Our economic wealth must be protected but we must also recognise that those riches were partly built through depleting the natural riches of our country, and as a result our generation has inherited a poorer America.

I have recently read again the words of Senator Robert Kennedy speaking in the University of Kansas, on March 19th, 1968. He said:

'Too much and for too long, we seemed to have surrendered personal excellence and community values in the mere accumulation of material things. Our Gross National Product, now, is over $800 billion dollars a year, but that Gross National Product – if we judge the United States of America by that – that Gross National Product counts air pollution and cigarette advertising, and ambulances to clear our highways of carnage.

It counts special locks for our doors and the jails for the people who break them. It counts the destruction of the redwood and the loss of our natural wonder in chaotic sprawl.

It counts napalm and counts nuclear warheads and armored cars for the police to fight the riots in our cities. It counts Whitman's rifle and Speck's knife, and the television programs which glorify violence in order to sell toys to our children.

Yet the gross national product does not allow for the health of our children, the quality of their education or the joy of their play. It does not include the beauty of our poetry or the strength of our marriages, the intelligence of our public debate or the integrity of our public officials.

It measures neither our wit nor our courage, neither our wisdom nor our learning, neither our compassion nor our devotion to our country, it measures everything in short, except that which makes life worthwhile. And it can tell us everything about America except why we are proud that we are Americans.'

Senator Kennedy was right, and his words ring true today. The US GDP today, despite the difficulties of the past six years, is around $15 trillion – nearly 200 times what it was in Senator Kennedy's day. Do we feel 200 times richer? Do we feel that our country is 200 times better? Are we 200 times happier? I say no, we are not.

Despite the economic difficulties of recent years, thanks to the measures the world put in place, and thanks, in part, to the economic measures that my administration has implemented, the GDP is rising again. I should, perhaps, be standing here and simply claiming the credit and making political points. But it is because we are now back on track that we must look down the track. Is this the path we wish to follow? I say no, it is not.

We are told that the consumption of renewable resources by the world's human population is 50% higher than that which the planet can sustain. You don't have to be a scientist, an economist, or a president to realize that is an untenable position for our species. But scientists and economists are telling me that, and I believe that they are telling the political leaders of all the world's countries the same message. I, as your president, have listened to that message. And all we Americans should listen to that message because we know, and it is an inconvenient truth (I use the phrase deliberately) that we here in America are amongst the very greatest of the world's consumers of the planet's resources. We know, in our heads and in our hearts, that we are taking more than our fair share and that by doing so we are short-changing future generations of Americans and present and future generations of our fellow world citizens. That is not a wise nor an honorable path to follow.

If we measure our progress solely by economic wealth then we are ignoring much of what it means to be an American and much of what it means to be an inhabitant of this planet. The United States is a powerful economic force, and with that power comes responsibility. We must ensure that our economic prosperity is not at the expense of our own, or others', social and environmental well-being. I commit the United States to being not just a military superpower, but an environmental one. We shall seek to make America's influence on the world one that defends the environment as well as defending liberty and freedom.

Previous presidents have done much for the American environment. President Lincoln laid the groundwork for the protection of Yosemite Valley and the Giant Sequoia in the latter years of the Civil War. President Theodore Roosevelt was a strong advocate for conservation, and under his presidency the

National Parks grew in number and extent and the Forest Service was formed. President Nixon's administration brought forward legislation to protect marine mammals, water and air quality, founded the Environmental Protection Agency, and brought in the Endangered Species Act – which has prevented further extinctions like that of the Passenger Pigeon. And, yes, of course I recognize that all these presidents were Republicans, but they were great American presidents and great Americans, though none was perfect – as no president is perfect. *[The president smiles, winningly.]*

America once led the world in its environmental thinking and action. Partly because of events like the extinction of the Passenger Pigeon and the near-extinction of the Bison, earlier generations brought in National Parks and legal protection for threatened species. But in recent years we have not led on these issues – we have lagged. We have not behaved as though the environmental health within our own borders, or that of other countries with whom we trade and exchange goods, is of concern to us. We have seemed to the rest of the world to be overly preoccupied with short-term material gain and not to pay attention to long-term environmental protection. And we have seemed to be environmental isolationists – only caring about our own short-term comfort and caring little for the impact that our consumption and trade has on the rest of the world.

Today I say to you that the environmental challenges we face are real. They are serious and they are many. They will not be met easily or in a short span of time. But know this, America – they will be met. It is time for the current generation of Americans, whether living in California or Maine, whether black or white, whether rich or poor, to make their contribution to protecting and enhancing the environmental quality of our great nation, and of the world as a whole.

And so I pledge that in the last years of my presidency, working with members of Congress of all parties (for the environment should be one of the least political issues of all), I will implement the following 10-point plan:

1. The United States of America will take account of a country's environmental record, as well as its record on human rights, in all our foreign relations.

2. The United States of America will ratify the Convention on Biological Diversity.

3. The United States of America will play a major role in international climate discussions and treat these as part of our role as a leading world nation.

4. We shall change our position within the World Trade Organization to argue that environmental considerations should be an integral part of international trade.

5. Domestically we will reduce our greenhouse gas emissions, by the end of the decade, by 10%, and by the end of the next decade by a further 10%.

6. We will remove public subsidies from biofuel production and increase them for truly renewable energy production.

7. We will create three new National Parks, whose role will be not only to protect existing natural resources but to restore and recreate some of the wilderness that we have lost. One shall be a forest National Park east of the Mississippi, another shall be an area of Tall-Grass Prairie and the third an area of Mixed-Grass Prairie. Funding will be made available for such environmental enhancement.

8. The budget of the National Park Service, the Fish and Wildlife Service, and the Environmental Protection Agency will be protected and modestly enhanced.

9. A new museum will be built on The Mall in Washington DC to celebrate the natural beauty and wildlife of the USA and America's role in global biodiversity protection.

10. September 1st each year will be a national holiday, named Environment Day.

Well, I guess that's a speech that will never be made. But if it were, what an impact it would have.

This is where this narrative leaves the USA and its missing Passenger Pigeons and heads home for the final chapter, as I did on the plane from Houston. It's almost time to say goodbye to the Passenger Pigeon.

The Passenger Pigeon was a species with a unique ecology, and it was uniquely numerous amongst birds. Its rush to extinction was uniquely great in terms of numbers of individuals lost from the planet and uniquely rapid amongst all avian extinctions. But it was not alone in its decline. Other American species were suffering too at the same time and for the same reasons. And since the Passenger Pigeon's extinction many more species of mammal, bird, amphibian, reptile, fish, invertebrate and plant have been driven to extinction. The Passenger Pigeon is a potent symbol of species loss, habitat loss and environmental loss – and the centenary of its extinction should prompt us to consider our past, present and future impact on the very thing that makes this planet special – the diversity of life.

We need not continue with the scale of ecological destruction, provided we change our ways. We can live lives that are rich in material wealth, with all sharing more equally in that wealth, and live on a planet which retains much of the natural beauty that still survives. If anything, given what a wonderful species we are, that sounds a little unambitious. But even to realise this ambition we will need to make a decision to change course – and quickly. That will require leadership.

So, from considering an extinct bird we have come to considering what we want from the future of human life on Earth and what constitutes Progress. Maybe Roger Tory Peterson was right when he wrote: 'The observation of birds leads inevitably to environmental awareness.'

Was that a tolling bell?

CHAPTER EIGHT

Bringing it all back home

> We must combine the toughness of the serpent with the softness of
> the dove, a tough mind and a tender heart.
>
> *Martin Luther King Jr*

I returned from the USA after more than five weeks of travelling
and writing in late June 2013. I had seen my first Blue Whale and
my first Giant Sequoia, I had seen the tall-grass prairie in Kansas
and the Saguaro desert in Arizona. It had been an inspiring and
useful journey, but it was good to reconnect with friends and
family, to be back home and to enjoy one of the best English
summers for a couple of decades. England retained the Ashes,
rather convincingly, against Australia, and my garden was visited
by many butterflies – neither of these things happens in the average
English summer. I settled down to finish writing this book.

My head was full of Marthas and pigeons. Only 12 hours
separated my being in the USA and the UK, but there was also the
gap of a century between the Marthas and me. As I settled back
into the familiarity of being in my own home rather than in a
different motel almost every night, and made arrangements to see
friends instead of being an Englishman abroad, my head was still in
the USA of the nineteenth century as well as in the USA and the
UK of the twenty-first. As I noticed familiar wildlife around me at
home I was thinking of clouds of Passenger Pigeons, progress and
how I wanted the world to be. Is there a message from Martha that
applies as much in Northamptonshire as it does in Ohio?

I live in a small town in east Northamptonshire. My house, red-
brick and semi-detached, was built in 1899, the year of the death
of the last wild Wisconsin Passenger Pigeon and the year before
Buttons was shot in Ohio by Press Clay Southworth. The street
where I live is quiet and residential. In the day it is largely empty

and in the evening it fills up with the cars of those returning from work in Wellingborough, Kettering, Peterborough and Northampton. Our street was not designed for two-car families, and the vehicles fill the roadsides and spill onto the pavement because this is England and it was here before the car was thought of. The view down my street often holds my gaze – the spire of St Peter's Church is beautiful and I am glad that it is there as a connection with all those who have lived in these parts over the years. The church was started in the twelfth century but its fifteenth-century wall paintings are what make it famous. The font is fourteenth-century. Yes, this is England.

But it reminds me of Ohio, and of Coshocton, Martha Grier's birthplace, where I stayed for a couple of nights, and of Bridgeport, where she spent most of her life and is buried. If an American visitor were looking to get a screw replaced in his sunglasses I hope he would get as friendly a reception as I did in Coshocton, and the town hall would probably be able to give an American visitor information on burial sites as efficiently as I received them in Bridgeport. You won't find the camera battery I needed on sale in any of the shops here, which is why I searched for it, unsuccessfully, in RadioShack, but you would probably get a similar friendly chat, and if you asked where to eat that evening you might be directed to the Mughal Dynasty Indian restaurant or the local Italian instead of The Warehouse. On the edge of town, symbolising the special relationship between our two great nations, there is a BP Garage where I get petrol but an American would get gas, and right next to it is a McDonald's where I could get a takeaway (but never do) and an American would find fast food.

The countryside around also reminds me of Ohio. There are rolling, undramatic hills with patches of woodland, mostly oaks and Ash, and the land is farmed. The farming is mostly for arable crops, wheat and oilseed rape, with very few livestock. There are more hedgerows here in Northamptonshire, and you may catch a glimpse of a distant church spire which would be a give-away – but, as with the towns, the countryside is noticeably similar.

However, the wildlife is different. My Peterson *Field Guide* wouldn't help me here – the only species of overlap are the House Sparrow and Starling – those European invaders to America (which thrive in the USA and decline in numbers here in Europe). There is no Passenger Pigeon to be missed from these skies, and no equivalent.

In each case the land used to be covered in deciduous forest, and the forest held wolves and bears. Here in Northamptonshire the wolves and bears disappeared centuries ago. Wolves had probably been extirpated from the whole of England by the time Europeans were discovering America, and a medieval artist was painting a mural – of Death spearing Pride with a lance – in my local church.

Near my Northamptonshire home, a landowner, in 1370, was given his land on condition that he found dogs to hunt wolves in Northamptonshire, Oxfordshire, Rutland and other counties. Bears disappeared long before that – probably before the Norman conquest of the eleventh century. But some Ohio woods held wolves until the 1850s and bears until the 1880s. Both Ohio and Northamptonshire lost their wolves, their bears and most of their forests, but the changes happened much more recently and much more quickly in Ohio as the spread of farming and the growing population made its mark on the American ecology.

I'm glad that Black Bears are spreading back into Ohio from West Virginia, and glad that they seem to be welcome, but my baselines have shifted so much that I cannot now regret the loss of Brown Bears from these local woodlands, nor can I wish them back. And I can stand outside my house and look towards St Peter's church spire and wonder how often a wolf passed along the ridge which is now my street, but I cannot imagine it back there, or wish it back there.

There is a much more recent, and much better documented, loss of abundance of wildlife in the area where I live. Some of it, the birds, I have noticed myself in the time I have lived in this area, but I haven't lived here long enough, and I don't have a good enough naturalist's eye, to have noticed the passing of plants and insects from my locality. However, in twenty-first-century England we do have a lot of information about the state of our wildlife over the last few decades – and those data paint a disturbing picture.

I am just one of 3,000 volunteers who collect bird data for the BTO, the RSPB and the Joint Nature Conservation Committee through their Breeding Bird Survey (BBS). I had surveyed my plot, which I started visiting in 2005, before I set off for the USA in May, and I just had time, within the rules of the scheme, to do the second visit and complete this year's data collection on my return.

30 June 2013

Data collection for the BBS has to be completed by the end of June, so this was my last chance – but only five days earlier I had been standing on the Congress Avenue Bridge in Austin, Texas, watching Mexican Free-tailed Bats! After a five-minute drive and a ten-minute walk I was at the point at a hedge that only I know is the place where I start counting birds. It was a fine sunny morning, but at 5:25 am there was little warmth in the air.

The task today, as it has been on 17 previous early mornings, is to walk along a set route of two kilometres in length and record the birds I see and hear in each of ten 200-metre lengths. My route is wholly on rights of way, a green lane and some footpaths, and sticks to the field margins, alongside hedgerows, some of which have occasional trees. The fields around me are all cultivated. They always contain either wheat or oilseed rape, and the only variation from year to year is which field has which crop. 'My square' – it's difficult not to feel a little possessive over it even though it has real owners who are farmers – is typical of the farmland for miles around. And that's what we would expect, as it is a one-kilometre square selected at random by some boffin in the BTO and allocated to me. The birder in me would like somewhere a bit more interesting; the scientist knows that this is the right way to do it.

It's an easy walk and I note down the birds as I go. Many are heard rather than seen, and I'm relieved that five weeks in the USA haven't removed the recognition skills from my brain. I couldn't do this survey in the USA – it would take me years to grow

American ears for American birds – but here it's easy to note down the Bullfinch from its quiet call and the Lesser Whitethroat that 'rattles' unseen in the hedge. Pheasants call noisily in the distance and Skylarks are calling and sometimes singing over the fields. It's quite easy, it's quite interesting, and it's quite normal.

If you get up at five in the morning then you feel you deserve a treat – at least I do. And nature often gives me one on these early mornings. One year a Montagu's Harrier flew through the survey area and made my day. Today it's a flock of four Curlews which passes overhead, unexpectedly, heading west. Those Curlews, like the harrier, won't contribute to the population change data that this survey measures, as they are just passing through – they are just a treat for me.

With the survey done by 6:30 am I head home, and over a cup of tea I enter the data online into the BTO database. I always do this as soon as I get home to get it out of the way. As I tap the keys I know that this completes my contribution to the BBS survey this year.

I take the time to look at the headline results from my square over the last nine years. It's quite a short time so I don't really expect to see much of any significance. Over the years I've seen 49 species in all, with 31 of those being present in 2005 and an overlapping 31 in 2013.

However, when I look at the primarily farmland species, then I do see some changes. There are 19 farmland birds which have been selected by government to provide an index of farmland birds, and, arguably, of farmland wildlife. The last Labour government used this index as a formal measure of sustainability, but when our current Conservative and Liberal Democrat coalition came into power they downgraded the importance of this indicator within months of taking office.

Of the 19 species that make up the index, six have never been recorded on my survey visits since 2005 (Grey Partridge, Kestrel, Turtle Dove, Rook, Corn Bunting and Tree Sparrow). Kestrel and Rook are slightly surprising omissions, as they both live locally, but they don't seem to use 'my' square very much. I'm not at all surprised that the other four species are absent – in fact I'd be delighted if I saw a single one of any of them, as they are species undergoing national declines. I'm fairly sure that they were all present here once, and maybe only 10 or 20 years before I started surveying this square.

For the other 13 species I do a simple sum as follows. I omit the middle year (2009) and then for each of the first and last four years I total the peak counts each year for each species. It's a simple and easy way to look at the data. I'm surprised to see that only one species has increased (Goldfinch, from a total of two in the early years to 14 in the later ones) and one species has remained steady (Reed Bunting, two and two). The other 11 species have declined as follows: Lapwing, nine and zero; Stock Dove, 11 and 10; Wood Pigeon, 64 and 43; Jackdaw, 13 and 2; Skylark, 49 and 40; Whitethroat, 23 and 19; Starling, one and zero; Yellow Wagtail, 16 and four; Greenfinch, eight and zero; Linnet, 11 and eight; Yellowhammer, 22 and 10.

I'm quite surprised at how gaunt these figures appear. I actually wouldn't have predicted them even though I know full well that this suite of birds is in trouble. The declines that I have recorded in a fairly brief (nine-year) period have been gradual enough for not even me, the observer, to have realised their full magnitude. It's a salutary reminder of the importance of data and the unreliability of memory.

I even catch myself thinking, 'but nothing has changed in those nine years' before giving myself a mental telling-off on two counts. First, although there has not been any major land-use change in that period, no houses built, no change of crops, no tree planting or removal, I don't really know enough about the area through which I have walked to say that nothing has changed. Maybe different pesticides are being used, or different amounts of them. There could have been all sorts of subtle changes that I would not have noticed.

But the second reason for my ticking myself off is that many farmers have said just the same thing to me – farming practices haven't changed over the last so-many years – and I have always replied that they are looking at things in the wrong way. If you go on a diet for two months you don't have to eat less and less every day for the diet to work. If you reduce your calories enough, and then keep your intake steady at that new lower level, you will probably still be losing weight at the end of the period. You wouldn't say, 'But I have been eating just the same for the last week, so it can't be anything to do with my diet' – at least I hope you wouldn't. British wildlife has been put on a starvation diet by British farming for decades.

You can't read much into the data from one small area of land over a nine-year period – but then we don't have to do that. We have data, from this scheme, for around 3,000 randomly selected squares covering farmland, woodland, urban areas and uplands, and they stretch back to when the scheme was established in 1995. And before that, going back to the early 1960s, there was a predecessor scheme with a smaller number of non-random sites that did a pretty good job for more than three decades.

As I look at the latest report on the BBS scheme, which covers 1995–2011, the fate of farmland birds in England is very clearly one of overall decline. Of the 19 species, 12 have declined between 1995 and 2011 and only seven have increased. And if we take our baseline back to 1970, just over a 40-year period, then 14 of the 19 species have declined and some of them were declining throughout this period. My observations for a few years in one small corner of east Northamptonshire form part of a picture of declining farmland birds over decades which affects England as a whole, and indeed the UK as a whole.

And it doesn't stop there. When I was in Nevada in late May, a group of wildlife NGOs and monitoring organisations were launching their *State of Nature* report back in London. I've read the report and I've seen the videos. The great Sir David Attenborough launched the report, and it shows that what has happened to birds has also happened to plants and invertebrates, even though the data aren't collected in anywhere like as systematic a way as the birds are surveyed, and even though they don't stretch as far back in comparable form. But then they don't have to in order to show dramatic declines. For example, the recent Plantlife report, *Our Vanishing Flora*, shows Northamptonshire as the county with the third-highest rate of loss of plant species across the UK. Every decade, slightly more than eight plant species are completely lost (extirpated – I do like that word, even though I don't like what it indicates) from the county.

This type of loss of abundance is different from the loss of the Passenger Pigeon. There are no global extinctions involved. The species we are considering are not being made extinct; they aren't even being extirpated from the UK, or from England, or (most of them) from Northamptonshire (though increasingly that is happening) – but they are being lost from our everyday lives. When, I wonder, was the last Grey Partridge on my survey square? And was 2013 the last year when I will record Yellow Wagtail – or is that still a few years away?

Each of the species on the farmland bird index has its story, but one stood out for me, and that was the Turtle Dove. It is the farmland bird showing the greatest decline since 1995 – an enormous 81% reduction in England. Is it going the way of the Passenger Pigeon?

19 July 2013

I set the alarm for 3 am and the usual happened – I slept fitfully and decided at 2 am that I might as well get up. As I drove through the east Northamptonshire lanes in the dark there seemed to be an awful lot of moths out. I swung onto the A14 and headed east and then south – towards Essex and Abberton Reservoir.

The roads were almost American in their straightness and lack of traffic. I thought back to driving through the former haunts of the Passenger Pigeon and about today's quarry – the Turtle Dove. I was meeting up with the RSPB team studying this fast-declining European bird of farmland. Was it going the way of the Passenger Pigeon? And what were the causes of the decline?

As a 12-year-old on holiday in East Anglia I remember we went through the Fens, that large flat area of peaty soils to the east of Cambridge that once was a huge wetland and now grows carrots, sugar beet and potatoes, and we saw lots and lots and lots of Turtle Doves.

Back home in Bristol and Somerset, in the west, we didn't have Turtle Doves and so this small fast-flying dove was a new bird for me. In areas near Wicken Fen they were everywhere – flying up from the roads, in the fields and on the telegraph-wires. This was akin to those experiences that a birder has when going abroad (and I had never been out of the UK at that time) – you see a new bird and study it carefully on first and second sighting, marvel at its beauty and learn its plumage and then realise you are going to see it every day many times a day and start to take it for granted.

That was what my first Turtle Dove experience was like. They are beautiful birds with backs and wings almost tortoiseshell in pattern – but their name comes from the purring *turrr-turrr-turrr* song of the male – the 'song of the turtle' from the *Song of Solomon* that will have perplexed non-birders often.

And in those days the song of the Turtle Dove was heard in this land, at least in the eastern parts of it, and right across most of Europe. These days, numbers are much depleted and its population graph looks like it is falling off a cliff. Before heading off to the USA in early May I had travelled across the Fens, spent time at Lakenheath Fen RSPB nature reserve, and spent a day at the RSPB reserve at Minsmere on the Suffolk coast. I saw and heard birds that I wouldn't have dreamt of seeing more than 40 years ago on my first visit to East Anglia – Common Cranes, Hobbies, Cetti's Warblers and Little Egrets, to name but a few – but despite being keen to see Turtle Doves, and passing through the very places where I had seen lots back in the early 1970s, the Turtle Dove eluded me.

Closer to home, my local bird-watching patch of Stanwick Lakes in the Nene Valley in Northamptonshire used to provide me with annual Turtle Dove sightings even though it was on the edge of the species' range. My records start in 2005, and in that year I am surprised to see that I saw up to six Turtle Doves in a day. They were still present through 2006–2008, though in noticeably decreasing numbers – but my many visits between 2009 and 2013 have produced not a single record.

Maybe I would be luckier today. Actually, luck wouldn't come into it, as I was meeting up with RSPB staff who were studying Turtle Doves and working with farmers to try to improve the lot of this farmland bird.

As I drove south on the M11, Cambridge was to my left and I recalled my days there as an undergraduate. I had helped survey a plot of land as part of the BTO Common Birds Census on the edge of Cambridge. I couldn't remember if it had had Turtle Doves then – there was a good chance it did, as at that time Turtle Doves had increased in numbers in the UK – but the plot ceased to be surveyed when this very motorway was built across its western edge. The first *Breeding Atlas* of British birds, published in 1976, the year I went up to university, but based on fieldwork carried out in 1968–1972, described this as a 'time of good population levels'. When it came to the publication of the next *Breeding Atlas*, in 1993, based on fieldwork in 1988–1991, the Turtle Dove had been lost from the western and northern edges of its range and was contracting towards what is now even more its heartland of East Anglia, Kent, Sussex and Hampshire. The Common Birds Census index had fallen by about 60%.

A little further on there was a signpost to Sandy, headquarters of the RSPB where I had worked for 25 years. Although I grew up in the west of England much of my life has been spent in these eastern arable areas rather than the dairy fields of home.

I had already passed a McDonald's or two, and a Texaco, when I stopped hoping for a coffee at the Stansted Services – my early start meant I had time for a much-needed caffeine fix. It was, of course, Starbucks that was open (open 24 hours), so I had a medium latte. The man who gave me my coffee did not tell me to have a nice day, or smile at me, or do anything to make me feel that he or Starbucks gave a damn about my enjoyment of their product – how I wish for American customer care sometimes.

Although many of the brand-names by the roads were American, the rolling English roads showed every sign of having been made by rolling English drunkards – twisting and turning through Coggeshall, Marks Tey, Birch and Layer de la Haye until they deposited me at the causeway at Abberton Reservoir 15 minutes early. I'm almost always early – if I could be given back all the minutes I have been early I could do great things in that time – but Jenny Dunn wasn't late, she arrived on the dot of 5 am.

Straight away she was pointing the antenna for the radio receiver at a clump of trees and getting the characteristic 'wheep wheep' that showed the bird, or at least its radio transmitter, was alive and well. We didn't see the Turtle Dove but he purrrred at us from his clump of trees.

Talking to Jenny, I realised this study had started in 2010, when I was the RSPB's conservation director, so I had a stake in it too. The Turtle Dove had been known to be declining for years, so why had the RSPB started so late? Was it incompetence on the part of the man who was conservation director at the time? Well, maybe, although the Turtle Dove was but one of a long line of bird species that had been declining in the farmland through which I had driven for two hours to reach this point. The RSPB, and others, had been throwing resources at declining farmland birds for a good three decades before the day when I stood listening to the 'wheep wheep'.

Through the day, Jenny and Tony Morris told me of the study and of the work to improve the lot of the Turtle Dove. Only by understanding the reasons for the species' decline can we have a decent chance of stopping that decline – if we want to stop it. There was nobody studying the Passenger Pigeon; there are lots studying the Turtle Dove – will they make a difference?

This RSPB team is not the first to study this bird. Several years ago, Stephen Browne, of the Game & Wildlife Conservation Trust (GWCT), studied Turtle Doves in East Anglia and discovered that their breeding success was nowhere near enough to keep the population going. Their breeding season is cut too short these days because there is not enough seed food for them to raise several broods successfully. In the 1960s Turtle Doves laid 2.9 clutches from which they raised 2.1 young to fledging age. In the 1990s, when Stephen carried out his study, each pair laid 1.6 clutches of eggs from which they raised 1.3 young – not enough for the population to remain steady. In fact, so few that the population should fall by 17% each year through the inadequacy of that figure. Jenny and Tony should find out more about the current situation and try to get a handle on how to turn things around.

The biology of the Turtle Dove is both similar to, and dissimilar from, that of the Passenger Pigeon. Turtle Doves are not colonial but are dispersed across the countryside (although maybe not strictly speaking territorial), they usually lay two eggs (of that we are sure), their nest is a flimsy little pile of twigs and the sexes share incubation, with the female doing the evening, night and early-morning shift and the male doing the middle of the day. Turtle Doves are pretty much entirely seed-eaters, consuming a variety of small seeds of farmland plants such as Fumitory, clovers and vetches.

They will make several nesting attempts during the season, though the radio-tracking that Jenny and others are doing suggests that the birds might move around between attempts – so even with the modern technology the details of the birds' nesting behaviour remain rather elusive. We visited several nests to check on progress; all were in thick vegetation that was easier for a Turtle Dove to penetrate from above than for us to negotiate at ground level. We walked, in the gathering heat, through face-high Stinging Nettles, through ditches and through thickets that clutched at our clothes. But we saw five nests, of which two were discoveries today. And at another site a male flew with two fledged young and then purrrred from an oak tree before soaring and then gliding down repeatedly in a display – he was full of the joys of summer.

Field studies of birds are not nine-to-five jobs. You have to be active when the birds are active – and birds get up early – and you have to make the very most of good weather and daylight because the birds' breeding season moves on inexorably and you can't catch up the work a few weeks later. It disrupts your social life and personal relationships as you follow unsocial hours and make the very most of the hours and minutes. It's also not a case of cranking the handle and the results come tumbling out – you need a naturalist's eye (and ear) (and Jenny and Tony both have that), a scientist's brain (ditto) and the character and energy and determination to keep going (ditto ditto). Sometimes you will take your scratched and aching body home feeling that you haven't made much progress – but on other days the insect bites and nettle stings don't hurt because you have made a step forward.

We drove through a small town on our way to check a nest behind some playing fields. We'd already seen two nests, one beside a busy road where if you crawled under the edge of the scrub and then peered upwards you could see a female Turtle Dove sitting on her two eggs on an apparently flimsy and insubstantial nest. Turtle Doves sit tight – we were only a few feet away but the female dove didn't budge an inch. We could almost have reached out and touched the cars that passed, but we couldn't be seen in the thick bushes. Across the road, an old red telephone box marked the location of the nest and a row of cottages pointed their windows towards us but their occupants were probably still sleeping. The other nest was by a quiet road through the countryside where a man was vaguely interested in what we were doing – but not very! Driving through the small town, it felt like it was time for lunch but it was barely 8 am – workers and schoolchildren were waiting for buses, most shops were not yet open, dogs were being walked and newspapers bought. For most, on this day that was going to be a scorcher, the day had hardly begun – and yet the Turtle Dove team had already got a lot of work under their belts.

Although when you study a species you tend, inevitably, to focus strongly on that species and begin to see the world through its eyes (assuming your study species has eyes), your study is embedded in a greater reality. There are colleagues and volunteers to coordinate and enthuse, landowners to contact and talk to, expense claim forms to be filled in, and data to be transcribed and analysed. Jenny took several calls on her mobile from other members of the team dispersed across East Anglia. We phoned a landowner, met a nice old lady and her rather suspicious largely deaf husband (I guess you don't get asked whether some people can look around your field for Turtle Doves every day of the week), and we met another lady with 10 cats who now knows she has a

Turtle Dove nesting in her field. But, as I know from past experience, the feeling that you are discovering nature's secrets, and the excitement of knowing more and more about a particular species as time goes on, is a heady one for people like us. Others become bankers.

Jenny and her team are working through the range of potential factors leading to a Turtle Dove population decline. They are evaluating potential causes from disease (lots of birds have bugs and parasites – but do they make a difference to survival and/or productivity, and where do the diseases come from – might imported Pheasant poults be one source?) and predation (losses of nests are high, but then they always have been in this species – it doesn't look like anything has changed). The most eye-catching threat is shooting by Mediterranean hunters – particularly the French and particularly in spring – but is this really a big enough mortality factor to be causing a decline of the Turtle Dove right across Europe? Perhaps something is going wrong on the wintering grounds to the south of the Sahara, or on important staging posts on their migration. As always, there are plenty of options, and they may all play a part, but Stephen Browne's study seemed to be on the right lines when he suggested that changes in agriculture were reducing the abundance of plant seeds for this bird.

Because they eat seeds through the year, and feed them to their young, Turtle Doves need there to be a chain of plants seeding through the spring and summer. One of the plants most associated with Turtle Doves is Fumitory. Most farmers would call it a weed, but it is quite pretty – if you look hard and with a kind eye – and Turtle Doves love it. The loss of Fumitory from farmland, thanks to changes in crop rotations and greater use of non-selective herbicides, has created a break in the chain of seed plants on which Turtle Doves depend if they are to produce enough successful broods each year to maintain their population.

Tony showed me a field plot, especially planted with a variety of seed-producing plants that Turtle Doves are known to eat, which is designed to be a food source throughout the spring and summer. Establishment and management of this mixture (Fumitory, Black

Medick, clovers and vetches) is being trialled by willing farmers, and it is hoped to add this as an option in government environment schemes fairly soon. That's the type of approach that helps farmers to help Turtle Doves.

Before I left, Tony and Jenny located the nest of a radio-tagged male bird. We had visited the area earlier, knowing that the bird was present. Radio-tags tell you a lot about where the bird is, but at the end of the day (it was actually mid-afternoon, and very hot) you just have to get hot and sweaty and dive into thickets to find nests. Because of the time of day we knew (or rather Jenny and Tony knew, and they told me!) that the male would be incubating if he and his female had a nest – and we thought that they did. There was a strong signal, which we gradually pinned down to the edge of a small unused field, but it took a good 15 minutes to be fairly sure which part of the thicket of Hawthorn, Elder, rose and bramble needed closer examination.

Tony and Jenny adopted slightly different strategies for diving into scrub on a baking hot day, I noticed. Tony took the 'wear several layers to reduce the impact of the thorns, nettles and scratches' approach and looked like a beetroot as he emerged now and again from the scrub, but Jenny adopted the fewer-clothes approach and looked cooler as she picked prickles and thorns out of her shoulders.

There were several false starts, looking in what later proved to be the wrong spot, and a lot of frustration as we failed to locate the small nest with its small bird and, presumably, small eggs. They were in there somewhere – and we were close to giving up when the nest was found. The male left his two eggs only as Tony's outstretched fingers were almost at the nest to examine its contents. Maybe it seems mean to disturb the bird, but we knew he would soon come back, and subsequent visits to check progress would be much quicker and would involve no hassle for the bird. Over the next few weeks the team would check this nest, and others, and build up a better picture of nesting success and how many times a season the different pairs nested.

As I bought Jenny and Tony a slice of cake and a cold drink to thank them for letting me, rather uselessly, tag along and see and hear how things were going, we ran through the potential factors which could lead to such a large population decline in the Turtle Dove. There is disease, competition from the increasing population of Collared Doves, predation from Magpies and other predators of nests, predation from Sparrowhawks and other predators of adults, shooting pressure on migration through Mediterranean countries, land-use changes on the wintering grounds, loss of stopover sites on migration and reduced breeding success due to lack of food on the breeding grounds. All of these might play a part, and it would be wrong to think there is a single answer, but to date the research that has been done shows that the last issue – lack of seeds to fuel the breeding success needed to maintain the population – is a completely sufficient explanation for the decline. It's not the trigger-happy Frenchman with a gun in Les Landes that we need to focus on, it's the nice farmer down the road in Essex who says he is trying to feed the world who requires our attention if we are to save the Turtle Dove.

Just as with the Passenger Pigeon, it is our definition of progress that we need to examine if we are to save nature. And just as with the Passenger Pigeon, we need to look at the bigger picture if we are to understand how we could redefine progress to put some ecology into the economics.

I was musing on these things, and enjoying the parlous state of Australian cricket, as I drove home through a mostly Turtle Dove-free countryside. My journey took me within sight, if I had craned my neck (which didn't seem like a good idea on the dual carriageway), of the RSPB's Hope Farm – where farmland birds, including in some years Turtle Doves, are doing well. I made a mental note to pop in sometime soon and see how they were getting on.

23 July 2013
I sat in the kitchen of the RSPB's Hope Farm with Ian Dillon and Derek Gruar – both long-serving RSPB staff whom I knew well. Most things hadn't changed a bit since my last visit – there was still

the offer of a cup of tea, a chocolate digestive biscuit (or two) and a readiness to talk about farmland birds – but there was a new coat of paint on the walls.

Ian is the fourth RSPB farm manager since the time in 1999 when the RSPB went into 'proper' arable farming. Across the RSPB's 140,000+ hectares of nature reserves there is a lot of land that is farmed, and some of it is farmed in pretty similar ways to how a 'real' farmer would carry out his work, but Hope Farm was bought not as a nature reserve but as a working farm, and the aim has always been to maintain its productive nature as far as the main crops of wheat and oilseed rape are concerned – but at the same time greatly to increase the yield of wildlife.

The 19 farmland bird species were the targets, not just the Turtle Dove. Just as meadow birds have declined in the USA – birds such as the Bobolinks I saw near Ithaca, and the meadowlarks which changed from Eastern to Western as I drove through Kansas – so have Europe's farmland birds. The UK has done a better job at driving the birds from its farmland than most other European countries, but it is a Europe-wide issue. Everywhere you go you will find that the birds of farmland have declined, whether they be Skylarks and Tree Sparrows in the UK, Crested Larks and Black-bellied Sandgrouse in Spain, or Rollers and Lapwings in Poland.

Modern farming is the culprit – when we squeeze more and more out of the land we squeeze out the wildlife. In Cambridgeshire, where we sat and talked, this farm and many others have lost their hedgerows (though some are being put back, as they are at Hope Farm), have lost their livestock (and are now all arable), and have benefited from the development of more powerful plant-killing and insect-killing chemicals. Wildlife struggles under these conditions. The birds show these effects well, and the data are very good, so we know quite a bit about our losses, but the plants and insects have haemorrhaged even more and many are absolutely at rock bottom. It's almost as though it's inevitable that producing higher profits and more food entails removing life from the countryside.

Some of us didn't want to believe this, and didn't believe it.

There was enough research around, some of it done by the RSPB but also by the BTO and the GWCT, to show that a slightly modified farming system would allow more wildlife to survive. Not necessarily to flourish, but at least to survive. And with that in mind the RSPB bought its farm, which it decided to call Hope Farm.

The story is told elsewhere in detail, but the project has certainly worked – wheat yields have generally increased and at least match the levels achieved on neighbouring farms, and yet farmland bird numbers are now much higher than in 1999 on this little island of bird-friendly, wildlife-friendly, but also food-friendly, farming. Skylark numbers have quadrupled in this period thanks to research led by the RSPB and the management practices put into place on the farm. It's pretty simple: just leave small bare patches in your wheat fields and the Skylarks will love them.

A mixture of nectar-rich margins, well-tended (rather than scalped) hedges and a few beetle banks, nest-boxes and a couple of new small water features have led to an explosion of life. A whole range of birds are doing much better at the top of the farmland ecological food web than they had been doing on this farm a mere dozen or so years earlier. Yellowhammers and Linnets have both done well, and Grey Partridges, Lapwings, Corn Buntings, Yellow Wagtails and Turtle Doves have all returned to breed in small numbers, and in some years.

As we talked in the kitchen I realised that Ian and Derek were a bit glum. It hadn't been the best year for birds last year, and this year wasn't looking so good either. And it had also been tough from a farming point of view – poor weather through autumn and winter and into what was supposed to be spring had knocked back crop establishment here as on many local farms, and spring wheat had been planted for the first time in years and years. I tried to cheer them up, while munching through their biscuits, by pointing out that bird numbers were still way above the levels when the farm had been bought even though, nationally, farmland birds continued to decline. Hope Farm was bucking the trend in a very big way. Being a naturalist, and being a farmer, you are at the mercy of the weather and you have good years and bad years – you have to take a long-term view.

I wasn't surprised when they told me that Turtle Doves hadn't nested at Hope Farm this year, and hadn't even been seen there this spring and summer. The occasional pairs that have nested in the past have always seemed like bonuses rather than the promise of a recovery. Maybe Hope Farm has seen the last of them? Maybe we will all soon see the last of them?

Hope Farm wasn't set up to demonstrate how to save the Turtle Dove – that seemed too big an ask on what is now the edge of the bird's range. As they say in the City when markets fall – don't try to catch a falling knife, you'll only get hurt. But Hope Farm had caught a few falling knives – Skylarks, Linnets and Yellowhammers in particular. It's very impressive to be able to demonstrate that hope is not lost and that solutions, practical solutions, are readily available.

As I drove away from Hope Farm I couldn't see through the hedge to look at the crops – the hedgerow was now too thick. I could remember when the farm was bought and the hedge was a shadow of its current stature. And Skylarks were singing over the wheat fields today. Despite Ian and Derek's slightly gloomy miens this was a place of hope for farmland wildlife.

But, driving home, the fields by the roadside were pretty birdless. I stopped once or twice but heard no more Skylarks. I was unlucky – they are out there, but far less abundant than they once were.

I'd admitted to Ian and Derek that the failure to persuade many farmers to follow the lead of Hope Farm was one of my biggest regrets from my days at the RSPB. I think I'd been too logical and rational – the fault of many a scientist. I had thought that once we discovered the reasons for the declines in farmland birds we could move on to researching solutions that would work for farmers. And that once we had developed solutions that produced more birds and could be adopted by modern farmers then we would be almost there. We got to that stage in the early 2000s and then I thought that the last piece of the jigsaw was put in place when the government was persuaded to include Skylark patches and a range of other wildlife-friendly options in the environmental scheme that English farmers could enter and get paid for doing a little more for wildlife. The take-up of the scheme was high but there were too many management options, and few farmers chose some of the newer ones such as Skylark plots. And so the research has been done and the money is available, but as I drive around the countryside I see very few little bare patches in cereal fields that would show that the farmer cares for Skylarks. They ought to be a badge of pride for farmers, and their absence is a black mark against the farming community.

Farmers are badly led by the National Farmers Union, in my opinion. But it's farmers who get to elect their leaders, so they have to take a share of any blame that is going. For many years the leadership of the NFU hasn't had an environmental bone in its corporate body. You might say that it's not the job of a farmers' union to speak up for Turtle Doves or Skylarks, or Cornflowers or Field Crickets – but why not? There could be a farmers' union that took the environment seriously and led the industry to make changes that increased its sustainability, but it certainly isn't the union that farmers have got, at the moment. When I left the RSPB I wrote that the NFU was an anti-environmental organisation, and I'd stand by that now.

If the NFU had cared at all about wildlife they would have promoted those aspects of the government schemes that would allow Turtle Doves, Skylarks and Grey Partridges to thrive, but they didn't, and so those species are not thriving. Precious little extra food has been produced, the incidence of hunger in the world has not decreased, and no farmer feels that he has made a fortune as a result – but farming as an industry has lost a golden opportunity to re-engineer its image.

I suspect if you asked the average person in the street – and I don't mean in the middle of London, my local high street in rural Northamptonshire would do – they would say that farmers are subsidy-junkies who drive around in new Range Rovers and don't have to work very hard. How unfair – or, at least, how partially out of date. But farmers lost the chance to get the RSPB and environmental groups on board by cold-shouldering the best options in the agri-environment schemes. Instead they carried on saying that farmers were doing a great job (some are – but that wouldn't be what the Turtle Dove, Skylark or Grey Partridge would say), that farmland birds hadn't declined (when decades of good evidence accepted by government and industry show they have), and that even if they had it was probably down to Magpies and Foxes (which, again, the evidence shows that it isn't).

If only farmers had been led in the right direction then, farmland bird numbers would be edging upwards, farmers could crow about it and wildlife conservationists would be acknowledging the progress that had been made. And it would all have been funded by the public through environmental grants, which instead of leading to that happy outcome have largely been yet another way of transferring money from the hard-pressed public to the largest landowners in the country.

I was getting myself down now, and I'd started the drive home feeling pleased that my visit might have cheered up Ian and Derek. But almost as I reached home I passed the land of one of my favourite local farmers. There are plenty of good farmers out there, and I'm lucky to have got to know one of what I hope will be a new breed – Duncan Farrington. And I remembered that many years ago I had found a Turtle Dove on Duncan's farm, much to his and my surprise, and much to his and my delight. I wonder whether it is still there? I'd better contact him and go and see.

25 July 2013
Duncan Farrington is the type of farmer I like, and the type of farmer of whom we need more. He is a businessman and an entrepreneur, he's keen to learn and he's keen to develop his farm and his business. He's the sort of guy who would succeed in lots of walks of life.

When I contacted him about looking for his Turtle Dove he agreed immediately and said he'd make me a cup of tea before I left so that I could tell him whether I had spotted one or not.

I woke before the alarm, as usual, and was surprised to find that it was raining, as I'd been sitting out in the garden late yesterday evening enjoying a glass of Rioja and the sound of swifts overhead. Should I go anyway or should I do some writing instead? I decided that I would have a look for a Turtle Dove, as I quite fancied a catch-up with Duncan anyway and I could buy some of his produce from him, too.

Duncan grows oilseed rape like many other farmers right across the UK, but he turns it into things that he can sell to you and me rather than just selling it as a commodity. Years ago he bought the kit that olive farmers use to produce cold-pressed olive oil and started producing Mellow Yellow cold-pressed rapeseed oil. He sells it in attractive bottles and now has a range of other products including – my favourites – mayonnaise and garlic mayonnaise. Duncan took the risk and put his money into an enterprise where he then had to market the produce, and he has done that very well. Others are now following Duncan into the market, but he was the first and he has the advantage of being the most established brand. I wish him well.

Duncan is also a LEAF farmer. LEAF is Linking Environment and Farming, which sounds rather good, particularly in days when farming seems too often to be divorced from the environment. LEAF farmers like Duncan need to have more of a voice in the farming community – but I guess he and people like him are too busy doing a good job and making a living.

LEAF provides a home for environmentally aware farmers who don't want to go as far as organic farming. Organic farming, which is pretty good for wildlife, and for the environment as a whole, only accounts for about 3% of UK agriculture, and that figure hasn't shifted much over the last few years. Having seen quite a few LEAF farms I am pretty sure that they are doing a good job, but I wish there were some scientific evidence to back up my belief that they are better than the average – the work just hasn't been done.

I first met Duncan quite a few years ago when doing a bird survey for the RSPB (I was conservation director at the time, but this was simply as a volunteer). The project involved getting birders (like me) to survey farms of farmers (like Duncan), with the aim of providing the farmer with a map of the farm showing where the birds were and which were the species of greatest interest. Duncan, because of the type of keen and open-minded farmer he is, was first in the queue, and as I drove past his farm twice a day on my way to and from work, and he was just in the next village, it was easy for me to volunteer.

I was surprised by what I found on Duncan's farm. It had, and has, plenty of Linnets and Yellowhammers but also a couple of pairs of Lapwings that year, and the star bird was the Turtle Dove. The Turtle Dove was probably nesting in the scrub on the abandoned railway line and the male purrrred at me, and at the world in general (but particularly at other Turtle Doves), on my early morning visits.

I was keen to see whether it was still around – although after several more years of declining population I didn't think the chances were high. I sat in the car, and the car sat in the rain, and with some windows down I listened for purrrring. These weren't ideal conditions but I could hear the far-carrying song of a Song Thrush, and nearer to me a Blackbird sang too. A Wren shouted its song at top-voice. There were a few Collared Doves calling and lots and lots of Wood Pigeons. When the rain eased I got out of the car and listened again. But I didn't hear the song of the Turtle before Duncan appeared and waved me over to have a cup of tea.

One of the reasons that Duncan and I get on, I think, is that I know a bit more about farming than most people (but not very much) and I'm interested in it, and Duncan knows a bit more about birds than most people (but not very much) but is interested in them. It means that neither of us ever tries to teach the other to suck eggs but we both ask quite a lot of questions and want to know the answers.

Duncan asked me why Turtle Doves were declining so I told him it was all down to nasty farmers and he laughed. He got his farmer's guide to weeds and looked up Fumitory and showed me the pictures. Rather than showing the attractive pink flowers of this nice-looking plant as the main photograph, that image was relegated to a small corner and pride of place was given to showing the emerging leaves of the plant so that any farmer would know what to do to stop it becoming a pretty flower.

As I had done with Ian and Derek a couple of days earlier, we chatted about the poor weather – which Duncan described as an 18-month winter – and what that meant for his crops. Rape is a resilient crop and often produces lots of seeds even if it looks half-grown and half-dead – Duncan was hoping that the harvest, which clearly wasn't going to be great, at least wouldn't be too bad.

I left with a couple of jars of mayonnaise, which Duncan wouldn't let me pay for (I told you he is the type of farmer I like), and I said I'd give his Turtle Doves another couple of chances some time when it wasn't raining as hard.

27 July 2013

I had another look for Turtle Doves at Duncan's farm – no luck. It had been raining overnight and everything was clean and wet. As I sat on a piece of old agricultural machinery by the line of the old railway, a clean, wet Fox trotted down the path towards me. I wondered whether it would see or sniff me and then turn and head back, but as I watched it out of the corner of my eye, so that I didn't scare it, I think I spotted the moment when it realised it was going to walk past me; it hardly paused at all but there was a moment when it held its foot in the air for a tad longer than I expected, and then it walked past. It seemed unperturbed, but my pulse quickened a little, as I always enjoy these close encounters with wildlife. I see Foxes more often on my visits to London than I do around my rural Northamptonshire home.

A distant Red Kite flew over the fields of oilseed rape. That would have been an exceptional sight when I first moved here in the late 1980s but now, if one puts in just a little effort, it is an everyday occurrence. Down the road a couple of miles, in my urban back garden, I expect to see a Red Kite when I look up, and I have been distracted from writing now and again by the cry of the kite over my shoulder as I sit facing a computer screen. Red Kites were driven out of most of the UK by persecution – mostly legal at the time, but persecution nonetheless – and held on by the very tips of their talons in mid Wales, thanks to their tenacity and the dedication of conservationists. There have been several Red Kite reintroduction projects across England and Scotland and Northern Ireland and now many more people enjoy the sight of these long-tailed, red-tailed, fork-tailed birds.

After about half an hour of listening to the Wrens blasting out their songs (how can such a small bird make such a big noise?), and the Song Thrush singing from the woodland, and the Wood Pigeons cooing from everywhere, I turned to leave and to get back to my writing. As I did so a Hobby flew through the treetops with remarkable speed. Another predator up early, it seemed!

No Turtle Dove today. I wondered whether this was a fool's errand. It was more likely than not that Duncan had lost his Turtle Dove some years ago and I was sitting in the damp with no hope of hearing the purrrr or seeing a quick small dove whizz past. Still, if you don't put in the effort to look and listen you certainly won't see wildlife, so although the computer keyboard was calling me back I didn't regret the time I spent outdoors this morning.

Some misguided 'real countrymen' would probably make the link between seeing a Fox, a Red Kite and a falcon – and no Turtle Dove – to suggest that it is 'all those predators' that are causing the reduction in farmland birds. Well – it stands to reason, doesn't it?

Predators can cause reductions in their prey numbers, and not just at the trivial level that the Rabbit population is decreased by one if that Fox I saw eats a Rabbit tomorrow, but real reductions in numbers. However, they are usually temporary reductions, and if you discount non-native introduced predators, which have wreaked havoc across the world's fauna, there are very few cases where natural predators and their natural prey do anything but rub along together in a kind of stand-off balance.

If Turtle Dove numbers were to increase dramatically one year (I wish!), perhaps just by chance, then the local Sparrowhawks would probably take a few more of them – they'd be more obvious and would probably have to feed further from cover and in more exposed places, and perhaps they would fly around more to find food, because they would be competing with each other harder for food and nesting sites. And that might mean that a higher proportion of nests would be taken by passing Foxes or Magpies too, because the 'extra' Turtle Doves would have to nest in less safe places as most of the best nest sites were already taken. But if the increase in Turtle Dove numbers were a matter of chance (perhaps a good breeding season and fewer French hunters being active at the right time of year) then their numbers would probably readjust on their own – regardless of any impact of predation – to the carrying capacity of the farmland where they live. Only if resources were raised, for example by lots of farmers providing the type of countryside Turtle Doves like, would numbers rise consistently. But the other side of the coin is that when lots of farmers take away from the countryside the things that Turtle Doves really need then their numbers fall consistently – and that's what is happening now. And if resources allowed Turtle Doves to find a higher overall population level then there would be a few more of them for Sparrowhawks to eat – although whether they would when there are so many bigger, juicier Wood Pigeons, and Collared Doves too, is a matter of conjecture. That's a real-life story of hawks and doves, a very simplified one I admit, but we have to realise that Sparrowhawks and Turtle Doves have been living together for millions of years – yes, the numbers of either might affect the numbers of the other but the balance of that relationship was sorted out thousands and thousands of years ago.

It's only when we intervene that things might get out of kilter. If we tidy up the countryside so that there are few of the thickets that I saw in Essex with tight-sitting Turtle Doves, or if we spray the fields so that Turtle Doves have to fly further and feed more distantly from cover, then predation rates may indeed go up – but if we blame the Sparrowhawk or the Fox then we are not looking squarely at our own role in this.

Those who seek to shift the blame for the decline in a whole range of farmland birds onto predators are usually either those who have something to gain by demonising predators, or those who have something to lose by accepting that they themselves may have played a role in these declines of wildlife – they are those who shoot and those who farm.

The shooting community has a completely different view of predators from nature conservationists (and yes, there are some conflicted individuals who are in both camps). If your sport depends on breeding up non-native birds, such as Pheasants, and then releasing them in ecologically enormous numbers (40 million each year) into the countryside a few weeks before shooting season starts, then any predator that takes a share of those birds is potentially reducing your 'bag' on your day's shooting. There is a bit of a conflict there, and the shooters seem very unwilling to treat losses to native predators as a part of their 'field' 'sport'. Field sports are not a necessary part of the countryside. They are carried out either for enjoyment (but then some small boys enjoy pulling wings off butterflies) or for profit (but then drug peddling and prostitution contribute money to the rural economy too).

And farming as a whole seems to have made a decision that it will deny any role in farmland bird declines and/or diminish their importance. I think this is a mistake. If only we had farmers' leaders working with scientists and nature conservationists to find a way forward rather than denying that there is any problem.

28 July 2013
Today was a family day, but my mind was brought back to Turtle Doves and the loss of farmland birds by an interview with the president of the National Farmers Union in the *Guardian* newspaper. He said that extreme weather, driven by climate change, is the 'biggest threat to British farming' and, 'Last summer was just a deluge and plant protection products [pesticides] were incredibly important to us even maintaining a pretty poor harvest: without them, there would have been nothing. When you have rain after rain after rain, the level of disease that grew up within the crop was absolutely out of this world.'

The NFU criticised the temporary ban on neonicotinoid pesticides (brought in to help to protect tumbling populations of bees, insects which farmers and other growers depend on, although not those who depend on wind-pollinated cereal crops), and stated that farmers needed to use lots of pesticides because climate change is making their life more difficult.

Do most farmers really agree, I wonder, that the solution to one environmental crisis – climate change – is to create more of another one – even more use of even more chemicals which reduce wildlife for all?

On wildlife the NFU president said, 'As I travel around, I see a fantastic British countryside and I do not accept that the countryside and environment is going to hell in a handcart.' This just doesn't wash when the *State of Nature* report showed that 60% of farmland wildlife species (whose trends were known) were declining. Just a couple of months after the publication of an authoritative report, accepted by a government minister on the day of its launch, the NFU seems to deny that there is even an issue to address. I find this astounding – but the NFU president went even further and said, 'This is provocative, but if our wildlife is where it is today in 20 years' time, I think that will be a pretty good achievement,' and 'If we are producing the same amount of food as we are now in 20 years' time, I think that we'll have a crisis.' I just wonder what the NFU thinks is wrong with the success of Hope Farm, which shows that you can have both.

This is the modern face of farming in the UK. It claims to be wearing the badge labelled Progress, but it feels entirely regressive to me. I'm glad that there are farmers like Duncan who are quietly getting on with farming in a softer way – which is in no way a soft and fuzzy way. Duncan is an entrepreneurial farmer who cares for his land. I doubt he'll ever be the president of the NFU.

2 August 2013
A much cooler day today and not raining either. Yesterday I had been in London and the temperature had been more than 30°C. At least I had seen a Peregrine close to St Paul's Cathedral – but no Mexican Free-tailed Bats streaming along the Thames.

On the train coming home I didn't see a single combine in the fields, and not a single harvested field – this is a very late harvest.

Because of the poor winter and spring everything is quite late – the spring flowers were late in coming, the migrant birds were all very late, and the crops are a few weeks behind their normal stage of growth.

This morning the Song Thrush was still singing as I sat listening for Turtle Doves – I think he must be an unpaired male, putting in the hours, singing his heart out, just in case a female appears to make his day (and his breeding season!). I feel a little bit that way about the Turtle Doves – I'm losing hope and it is beginning to feel like a fool's errand. However, sitting by the old railway line is beginning to be my place of contemplation over what's wrong with the world.

Recently George Monbiot's new book, *Feral*, has been stirring things up. George, in his own engagingly provocative way, has been promoting the idea that we need more wild nature in our lives – he's right, we do. George is promoting the concept of re-wilding – putting the wild back into the countryside with unmanaged tracts of woodland, a few reintroduced top predators and a reduction in human input. I certainly support that, and the uplands of Britain would be where we should start. Let's welcome those Eurasian Beavers back, and some Grey Wolves, and the Eurasian Lynx too.

But I doubt whether we will be welcoming them back to rural Northamptonshire that quickly. The area over which I look, and over which a Red Kite is calling although it is hidden from view, did have all those species, but a very, very long time ago. A couple of thousand years ago I would have been standing in a forest that stretched beyond the horizon in every direction, and the open vistas would have been the river valleys, like that of the River Nene just a few kilometres away (and we are not talking of a Mississippi here – it's a British river, a homely river, not a mighty river), and areas where lightning fires caused gaps in the forest for a few years.

A Jay flew over squawking. The Jay is one of the few nest predators who is sneaky enough (in other words clever enough) to find a Turtle Dove nest – I wonder whether Duncan's Turtle Doves disappeared inside a Jay? Jays are woodland birds which have adapted to the sparsely wooded open field-scape of much of modern Britain. And the Jay is making its living for some of the year rather like a Passenger Pigeon – in the autumn it feeds largely on acorns, beech mast and chestnuts. We see Jays carrying acorns in the autumn to bury them for retrieval later in the year. The Jay has proved able to cope with the mixed landscape of Britain in a way that the Passenger Pigeon could not cope with the rather similar landscape of Ohio.

I doubt whether Duncan would be thrilled at the idea of re-wilding his land. How would he make any money out of it? Would the state pay him not to farm, as we now pay him, through subsidies, to be a farmer? And if it did, would this appeal to him? My guess – a strong expectation actually – would be that it wouldn't.

Today the papers carried reports based on the Breeding Bird Survey up until 2012 that the Turtle Dove has fallen in numbers in England by 85% since 1995 – one more year's data shows a slightly greater decline. Now the population is only around a seventh of what it was 17 years ago. And the press release from the RSPB suggests that this year is a worse year still. I can't help thinking of Jenny and Derek and their painstaking work to find solutions, of Duncan and the fact that I couldn't find a Turtle Dove on his farm, and of the NFU's lack of sympathy for wildlife.

16, 17 and 18 August 2013
I took three days off writing to go to the annual Bird Fair at Rutland Water. It's sometimes billed as 'Glastonbury for birders' – I think I saw Mick Jagger in the distance – or was it Derek Moore? And was that Charlie Watts or Nigel Redman, Ronnie Wood or Martin Davies, Keith Richards or Dick Newell? The smell of trampled grass, alternating downpours and hot sun, decent beer and fast food (once you've stood in a queue for a while) would certainly make some festival goers feel at home.

The Bird Fair has its celebrities too. You may spot Chris Packham, Bill Oddie, Nick Baker, Simon King, Mike Dilger or David Lindo mixing with the masses. If you haven't been to a Bird

Fair then do think about going. Some go for the excellent talks –
but I hardly ever get to a talk. Some go to choose their next foreign
holiday with birds at its focus – but I've never bought a holiday at
the Bird Fair. Some go to replace their binoculars or telescope –
but I've never done that either. I go to the Bird Fair to meet my
mates!

And this year we were talking about all the usual issues and how
they were playing out: which wildlife NGOs were up and which
were down, the poor environmental literacy of the UK
governments in general and the current ministers at Defra in
particular, but most of all we talked about birds. We swapped
stories about what we had seen and what we hadn't. The plight of
the much-persecuted Hen Harrier was talked about because of the
proximity of the event to the so-called Glorious Twelfth – the start
of the red grouse shooting season. This year only two pairs of Hen
Harriers attempted to nest in England, when scientific estimates
suggest that in the absence of illegal persecution there should be
around 300 pairs.

We also talked about farmland birds in general, and the plight
of the Turtle Dove in particular. Things are bad, very bad, for
these species – but at least we can see that there is quite a lot of
action. The RSPB is working with those elements, sadly too few,
of the farming community who really care about wildlife to try to
make things better. And a car-full of RSPB ex-colleagues of mine
had seen a Turtle Dove in Bedfordshire on their drive north – a
happy event worth mentioning and celebrating, but which once,
not so long ago, would have been a perfectly normal and quite
unremarkable occurrence.

By chance, and not completely unexpectedly, I met three
farmers at the Bird Fair who fitted in well (and two of them were
mentioned in my earlier book, *Fighting for Birds*, as being some of
the good guys). Nicholas Watts (MBE no less) farms in Lincolnshire,
in the Fens near Market Deeping. Vine House Farm was one of
those used by Stephen Browne in his Turtle Dove study, but despite
Nicholas's best efforts at wildlife-friendly farming, numbers have
declined. Nicholas knows his birds and would probably be at the

Bird Fair even if he didn't have a large stand selling bird food to us birders – for he is a birder too. Some of his land is farmed organically, and Vine House Farm has won the prestigious Farming and Wildlife Advisory Group Silver Lapwing Award twice – the only farm to do so – in 2001 and 2011. I wished Nicholas well as a finalist in the equally prestigious (if not more so?) 2013 Nature of Farming Award run by the RSPB, Butterfly Conservation and Plantlife (with the *Daily Telegraph*).

My second farming friend was Rob Law, who farms on the light land near Royston on the Cambridgeshire/Hertfordshire border. He said that he could only come to the Bird Fair because it had rained yesterday and harvest was put back a bit. Rob supplies Conservation Grade cereals to Jordans – so you may be eating his produce for your breakfast each morning.

Patrick Barker tapped me on the shoulder and gave me a big smile. He and his father were heading off to a talk so there wasn't time to chat – unfortunately. This is another FWAG Silver Lapwing Award winner – for 2009. Patrick and his cousin Brian – known as the (fabulous?) Barker Boys – were named by *Farmers Weekly* as farmers of the year in 2010. They have Turtle Doves on their farm, and the RSPB has done some work there too, but their numbers are dropping as well. I wish we could have talked – but we were heading in opposite directions and it was nice that we had, at least, smiled at each other and grabbed a few quick, friendly, words.

It would be good to see more wildlife-friendly farmers at the Bird Fair and to hear what they think about the state of nature in the countryside, and also to hear about the good things they are doing. But in the absence of such events, it is good to see some leading farmers attend the Bird Fair every year – and attend it as willing and enthusiastic participants rather than under any sense of duty. They come because they like it – and they care about birds and wildlife too.

That reminds me – I must go and have another chat with Duncan.

20 August 2013

I went to have a chat with Duncan just to confirm that I'd failed to find his Turtle Dove this year. Duncan was busy with the harvest. Tomorrow he hoped to finish the rape harvest (a couple of weeks later than a normal year) and he described it as 'disastrous to not good'. The wheat he was harvesting was of good quality but of low yield, and the prices were low on the global market. He had sown spring barley for the first time since he was a lad, and it was looking really good, but Duncan was wondering whether there might be a glut of seed barley – as everyone else might have done the same as him. Still, he looked pretty cheerful considering.

As I left Duncan, wishing him the best of luck with the last of the harvest, we agreed that I'd come and have a proper search for his Turtle Doves next spring. As I drove away I thought that it was unlikely that there would be good news on Turtle Doves next year but I would certainly have a look and a listen. I know Duncan would be pleased if I found him a Turtle Dove, and I know that I would be too – at least we can have another cup of tea and exchange thoughts on farming and farmland wildlife. And I'll go back to survey my BBS square and hope that farmland birds will have increased dramatically.

Northamptonshire looks a little like Ohio, but even if it didn't there are parallels to draw between the USA at the time of the Passenger Pigeon's extinction and the present-day UK. When the Passenger Pigeon was driven to extinction, the world, and the eastern USA (and a part of southeast Canada) lost a few billion birds. According to *The State of the UK's Birds 2012* (a report published by five government agencies along with the BTO, the RSPB and the Wildfowl and Wetlands Trust), the UK lost 44 million breeding birds between 1966 and 2012. That's about one million birds a year on average, and it would represent a rate of loss of 100 million birds each century just for the UK. Extrapolated over the whole of Europe, this might easily add up to the loss of a billion birds in a century. And in terms of percentages, in the period between 1966 and 2012 the UK lost about 20% of its breeding bird population. Yes, some species increased in that period, but the net change was a 20% loss – which represents a loss, over a century, of slightly more than 40% of the UK's breeding birds. That's about the same, in relative terms, as the loss of the Passenger Pigeon from the avifauna of North America.

These are huge losses, and much of that loss is of the farmland birds that used to be the bread-and-butter wildlife encounters in the English countryside – the Skylarks, the Tree Sparrows, the Yellowhammers and, yes, the Turtle Doves.

There is no European farmland bird whose global extinction is imminent, but there are an awful lot of local extirpations already in the bag, and the UK extirpation of the Turtle Dove now seems more likely than unlikely – maybe within the next decade. Are we going to wait until the Turtle Dove disappears from the UK before we wake up and act? Isn't an 85% decline enough of a loss and a loud-enough tolling bell?

It's ironic that a dove is now leading the way in UK farmland bird declines as we mark the centenary of the extinction of the Passenger Pigeon. We ought to take the hint, even if we still are not completely sure of the cause of this species' decline. Maybe disease plays a part, certainly illegal spring shooting plays a part (although I would guess a small one), but it does look as though unsustainable agriculture is most to blame. If we were considering the Turtle Dove alone then maybe we should do more research before acting, but it very clearly isn't just the Turtle Dove. A whole range of farmland birds (and plants and insects) is declining.

We are in a different position from those who witnessed the disappearance of the Passenger Pigeon. Each year we add another point to the graph of Turtle Dove decline, and we get an update on the status of farmland birds as a whole. No such graph existed for the Passenger Pigeon. Its nomadic way of life, and the lack of monitoring schemes at the time when it lived, meant that it was far more difficult to spot what was happening – but we do not have that excuse. We are not in ignorance today. We know our wildlife is disappearing.

And we cannot say we are ignorant of the solution either. Hope Farm and many other examples show that wildlife declines do not need to accompany efficient food production. We can have our cake and our Skylarks – we do not have to choose one or the other. The work of the RSPB at Hope Farm and the efforts of farmers like Duncan, Nicholas, Rob and the 'Barker Boys' show that there is an alternative. It's that alternative I want, please. I'd like to be proud of British farming instead of ashamed of it. Why on earth have we not tweaked our farming practice to deliver more wildlife? It isn't because we do not know how to do it – it is because we have not, as a society and a country, shown that we care enough.

And despite the great efforts of many 'good' farmers, like the few I have mentioned here, it saddens and maddens me that farming has not responded to the challenge. In fact it has seemed to turn its back on the challenge of producing food in an environmentally friendly way.

I know I don't want the future offered to me by the National Farmers Union. Their offer, it seems to me, is a version of 'caring technology' – except I don't believe they care and I don't believe the technology will work. And if it does, it will certainly be at the expense of wildlife and natural beauty in all our lives. I'm really unconvinced that more GM crops, more and stronger pesticides and more land given over to biofuels are going to create a future in which I want to live. They will almost certainly, without extreme care, deliver a future in which even less wildlife will live. That's not what I want and it's not what I think we could have.

I would love to be in a position where I could feel pride that British science had led the way to British agriculture policy restoring much British wildlife whilst enabling British farming to be profitable economically and productive in terms of food.

When I look at my ability to influence these issues I feel a bit helpless, but I don't like feeling helpless, so let's find a way out. When I consider my position compared with that of Martha Grier I am clearly very powerful. I have much more information than she did about the consequences of my actions and those of others. Not only do we know more, but modern technology means that we can all become experts on anything pretty quickly if we really want to be. I am also much better off in terms of time and money than she can have been. This means that I can decide whether to spend a bit more on food in order to influence the markets – I can exercise my consumer choice. I also have a vote, which Martha Grier never had. Because £3.3 billion of taxpayers' money is given to British farmers every year I should be able to influence the politicians to spend that money much better – even if only about £55 of it is mine. And also through political influence I can help ensure that regulation plays a constructive part in creating the type of countryside I want to see – one that provides us all with safe and healthy food, farmers with a decent living and wildlife with the ability to thrive.

But these are tricky issues. I can only vote for change if I am offered it – and environmental issues are never that high on the agendas of the political parties because they, rightly, surmise that not enough of us care about them. I'll do my bit to raise the issues, in my daily blog (www.markavery.info/blog), the articles I write and the talks I give – but we really need an army. You could be part of that army by lobbying your political representatives (your MP in Westminster, your MSP in Edinburgh, your AM in Cardiff, your MLA in Belfast, your MEP in Strasbourg and Brussels, your local councillors) to do a better job for nature. Raise your voice and show that you care!

When it comes to my purchasing power then I am already doing some of the right things. I have switched to a green – I think the greenest – electricity supplier, Ecotricity, even though it will cost me a little more money (I'm hoping it will cost the planet less). And I have switched to paying for my metered water use rather than a fixed charge, as that may save me money – and anyway it will make me think about how much I use. There are solar panels on the roof and the house is better insulated since we replaced the windows last summer. I only eat meat two or three days a week to reduce my carbon footprint and I've gone on a diet too to lose inches and literally have a lighter footprint. I have hardly used a plane in the last decade (I know I flew to the USA twice but that's almost all of it), and my car use has dropped too.

However, buying food is a tricky area. How does one choose the green option that is fair trade too? Increasingly I am going to look out for the LEAF Marque products, produced by farmers like Duncan, and hope that supports the type of farming I want to see. It's not enough, but it's a start.

I'm going away to think about it some more and I will come up with a menu of pledges for you to sign up to if you want the type of countryside that I want too. By the time you are reading this book, I will have published the list of pledges on my website (www. markavery.info), and I invite you to show your support by signing up. If there are only ten of us we will make a small difference – but a difference nonetheless. If there are a hundred, then we can make a bigger difference. Imagine if we were a thousand – then we have the makings of a movement. Will you join me?

1 September 2013
I got to Crewe early – I get everywhere early! Crewe was a compromise halfway house on my journey to Liverpool. I would have preferred to go by train but the timings of the Sunday services, and their prices, meant that the car was a cheaper and more convenient (though less green) option. Parking at Crewe and getting the train into Lime Street Station seemed the type of decent compromise that I should more often make. I arrived after a two-hour drive before 9 am and thought I would get a train after 9:30 am. The 10:30 am was the next sensible option so I had time to kill. Even its best friend might admit that the environs of Crewe Station on a Sunday morning with grey clouds filling the Cheshire sky do not provide the most lively or inviting place to while away an hour and a half.

The Royal Hotel looked as though it had seen better days and wasn't looking forward with much optimism to the future, but I found a warm welcome and they were pleased enough to give me a cooked breakfast for £6.50. The woman asked whether I would like tea or coffee and I opted for the former – but there wasn't the long interrogation that would have accompanied a US breakfast order. I didn't have to choose between white, wheat, rye or sourdough toast and I didn't have to specify how many eggs and how precisely I would most enjoy them, but when the white sliced toast and the sausage, bacon, fried egg, mushrooms and baked beans arrived they were just what I wanted. The friendly staff brought me more toast, and were happy for me to linger for an hour over my breakfast.

On the slightly intrusive TV, the news was about President Obama's decision to consult Congress on the next step in Syria after David Cameron's government's defeat in the House of Commons a few days earlier. The French government announced that they weren't going into Syria on their own, which was hardly the most surprising announcement. This mixture of news from the UK, the USA and France seemed fitting for the day. I was heading to Liverpool on this, the 99th anniversary of the extinction of the Passenger Pigeon, to see the work of John James Audubon.

The refurbished Liverpool Central Library is four minutes' walk from Lime Street Station and hosts one of the 120 or so complete sets of Audubon's *Birds of America* in the original double-elephant-sized edition. It is the only one on display in the UK, and it was almost lost when the old library suffered a direct hit from a German bomb in 1941, and the strongroom where the book was kept was filling up with water from the firemen's attempts to quell the flames. A quick-witted librarian rescued the volume – and I imagine it must have been a big strong man, as a volume of 435 plates illustrating 1050 individual birds would be getting on for 25 centimetres thick, and it is a metre by 70 centimetres in size. I imagine him unlocking doors and cabinets, up to his knees in the rising water, then having to carry this massive book to safety. It is now on permanent display in the Oak Room on the first floor of the Library.

I had contacted the Library about a month earlier and pointed out to them the significance of this day for the Passenger Pigeon and asked whether the book, in its glass case, could be opened to

Plate 62 on this day to mark the occasion. After a couple of days'
thought the answer had come back 'yes' – and that was why I was
heading to Liverpool.

If the environs of Crewe Station can be described as
unprepossessing (and they can), then those of Lime Street Station
can best be described as imposing. The solid Victorian architecture
of the Walker Art Gallery (built 1877), World Museum (1860)
and St George's Hall (opened 1854) speaks of wealth and power,
and no doubt so would the mounted Queen Victoria, which I
passed, if a statue could speak.

Audubon had come to Liverpool in 1826 with sets of his artwork
to gather subscriptions for *The Birds of America*, and he had been a
great success. Dressed deliberately, we can imagine, in buckskin
with his long hair slicked with bear grease, he brought tales of the
frontier into the drawing rooms of the English rich and powerful,
and the most beautiful images of strange wildlife too. I'm glad that
Liverpool displays his work – it seems fitting.

And I may have played a small part in it being displayed. When
Liverpool was named European City of Culture for 2008 a small
group of us, led by the environment editor of the *Independent*
newspaper, Michael McCarthy, who hails from these parts, the
environment minister Elliot Morley, the journalist Duff Hart
Davis and myself, then the RSPB's conservation director, lobbied
unsuccessfully for Audubon's book to find a permanent display.
We were unsuccessful but maybe – I don't know (as is often the
case with advocacy, although it usually doesn't stop people claiming
the credit) – our words stuck in someone's mind, and when the
refurbished Central Library was opened *The Birds of America* was
put on permanent display in the Oak Room.

The new Central Library was busy this Sunday morning.
Families looked for reading books, students used computer
terminals and tourists visited exhibitions. It was just how a library
should be – busy!

I asked directions to the Oak Room and was greeted with smiles
by the helpful staff. A man took me to where I wanted to be and as
we approached I wondered what species would be displayed on the

open pages of this enormous book. Had they remembered that today was a significant day for the species on Plate 62?

They had! I thanked my guide with a few words about why this book, and this bird, meant so much to me and he seemed pleased but slightly bemused. I stood and looked at the image of a pair of Passenger Pigeons. The male stands on a lower branch, his graduated tale fanned, and the duller-plumaged female reaches into his open bill from the branch above.

Let's set aside the biological inaccuracy of the birds' pose and just drink in the beauty of the art and the beauty of the bird. Two life-sized Passenger Pigeons illustrated by a man who had seen billions of them. The most numerous bird on Earth, on Plate 62 of the most expensive book on Earth. And this day was the 99th anniversary of that species' extinction, the 99th anniversary of the day when Martha died in Cincinnati Zoo.

I was glad to see that the colours were lifelike, as far as I could tell from the skins and mounted specimens I had seen, and were not at all gaudy as they sometimes appear in reproductions of this plate. This was the real bird made beautiful on the page.

I spent most of the next hour in the imposing circular reading room nearby, looking through the Natural History Museum edition of *The Birds of America* – not as large as its original source sitting in the display case in the next room but easier to peruse. I started at the beginning and looked through the plates in the order in which they were produced and sent out to subscribers.

As I turned the pages I was reminded of birds that I had seen on my travels in the USA, but also of some that I had not seen, and could not now see. Plate 26 illustrates the Carolina Parakeet in all its lost glory; Plate 62 is the Passenger Pigeon; the Ivory-billed Woodpecker is resplendent on Plate 66; Bachman's pretty warbler crops up on Plate 185; the Pinnated Grouse, or Greater Prairie Chicken, includes the departed Heath Hen (Plate 186); soon after it comes the Eskimo Curlew (Plate 208), and then there is quite a gap until we meet the Labrador Duck (Plate 332) – and last of the 'friends of Martha' is the Great Auk on Plate 341. That's quite a roll call of extinct and probably-extinct birds – all laid low in a short period of time.

I closed the modern edition and went back to the glass case in the Oak Room. I looked deep into the eye of the female Passenger Pigeon, which had been one of billions when Audubon had painted its portrait – and I thought of Martha. If Martha could speak to us now then perhaps her message would be as follows:

I forgive you for wiping out my species – you didn't really mean to do it, and maybe you knew no better. I forgive you for all the other accompanying ecological destruction of those times – the same excuses apply. However, the excuses are slipping away. You can now choose what type of world you live in, and what type of world you create, in a way that no other species can. You can choose the level of future ecological devastation, and the excuse of ignorance no longer holds. Whether you do better in the future is a test of your worth as a species. You have the knowledge and ability to live sustainably on this planet but it's a hard road from where you are now. It's no longer a matter of what you know – you know enough. From here on, it's a test of whether you care – do you care enough? Please care. Please do better. Please start now.

Further reading

This is not a complete reference list of sources used in the writing of this book – I can't imagine many people would want such a thing. In the text I have tried to give enough information (even if it is only a forename, surname and indication of date) to enable the determined reader to track down the references on the internet.

Because much of the focus of this book is long ago, many of the sources are now out of copyright and are available on the internet. This applies to many of the scientific papers, including those published in *The Auk* (the journal of the American Ornithologists' Union), which can be found through the Searchable Ornithological Research Archive (SORA) at https://sora.unm.edu/node.

Similarly, although buying an original edition of Audubon's *Birds of America* is beyond most of our pockets, it is possible to see the plates (and read the text from his *Ornithological Biography*) online at http://digital.library.pitt.edu/a/audubon.

If anyone has a good reason, and a burning desire, to track down some references on Passenger Pigeons or other subjects mentioned in this book then please get in touch with me at mark@ markavery.info and I will do my best, other commitments allowing, to help.

GENERAL

Cocker, M. and Mabey, R. 2005. *Birds Britannica*. Chatto & Windus, London.

Cocker, M. and Tipling, D. 2012. *Birds and People*. Jonathan Cape, London.

Fuller, E. 1987. *Extinct Birds*. Viking/Rainbird, London.

Goodwin, D. 1983. *Pigeons and Doves of the World*. Cornell University Press, Ithaca.

Gibb, D., Barnes, E. and Cox, J. 2001. *Pigeons and Doves: a Guide to the Pigeons and Doves of the World*. Pica Press, Robertsbridge, Sussex.

Hume, J. P. and Walters, M. 2012. *Extinct Birds*. T & AD Poyser, London.

Schorger, A. W. 1955. *The Passenger Pigeon: its Natural History and Extinction*. University of Oklahoma Press, Norman.

Steadman, R. & Levy, C. 2012. *Extinct Boids*. Bloomsbury, London.

CHAPTER 2

Brewster, W. 1889. The present status of the Wild Pigeon (*Ectopistes migratorius*) as a bird of the United States, with some notes on its habits. *Auk* **6**, 285–291.

Cooper, J. F. 1845. *The Chain Bearer*. Burgess & Stringer, New York.

Flannery, T. 2001. *The Eternal Frontier: an Ecological History of North America and its Peoples*. Heinemann, London.

Kalm, P. 1911 (reprinted from the original of 1754). A description of the Wild Pigeons which visit the southern English colonies in North America, during certain years, in incredible multitudes. *Auk* **28**, 53–66.

Muir, J. 1913. *The Story of my Boyhood and Youth*. Houghton Mifflin, Boston.

Peterken, G. F. 1996. *Natural Woodland: Ecology and Conservation in Northern Temperate Regions*. Cambridge University Press, Cambridge.

Peterson, R. T. 1980. *A Field Guide to the Birds East of the Rockies*. Houghton Mifflin, Boston.

Price, J. 1999. *Flight Maps: Adventures with Nature in Modern America*. Basic Books, New York.

Westmoreland, D., Best, L. B. and Blockstein, D. E. 1986. Multiple brooding as a reproductive strategy: time-conserving adaptations in Mourning Doves. *Auk* **103**, 196–203.

Wilson, A. 1808–1814. *American Ornithology; or, the Natural History of the Birds of the United States: Illustrated with Plates Engraved and*

Colored from Original drawings taken from Nature. Porter & Coates, Philadelphia.

CHAPTER 3

Ellsworth, J. W. and McComb, B. C. 2003. Potential effects of Passenger Pigeon flocks on the structure and composition of presettlement forest of eastern North America. *Conservation Biology* **17**, 1548–1558.

Jaeger, M. M., Bruggers, R. L., Johns, B. E. and Erickson, W. A. 1986. Evidence of itinerant breeding of the Red-billed Quelea *Quelea quelea* in the Ethiopian Rift Valley. *Ibis* **128**, 469–482.

King, W. R. 1866. *The Naturalist and Sportsman in Canada or Notes on the Natural History of the Game, Game Birds and Fish of that Country.* Hurst & Blackett, London.

McGee, W. J. 1910. Notes on the Passenger Pigeon. *Science* **32**, 958–964.

Nilsson, S., Niklasson, M., Hedin, J. *et al.* 2002. Densities of large living and dead trees in old-growth temperate and boreal forests. *Forest Ecology and Management* **161**, 189–204.

Revoil, B.-H. (translated W. H. Davenport). 1874. *The Hunter and Trapper in North America or Romantic Adventures in Field and Forest.* Nelson & Sons, Edinburgh and New York.

Schorger, A. W. 1937. The great Wisconsin Passenger Pigeon nesting in 1871. *Proceedings of the Linnaean Society of New York* **48**, 1–26

Ward, P. 1971. The migration patterns of *Quelea quelea* in Africa. *Ibis* **113**, 275–297.

CHAPTER 4

Cokinos, C. 2000. *Hope is the Thing with Feathers: a Personal Chronicle of Vanished Birds.* Jeremy Tarcher/Penguin, New York.

Darling, F. F. 1956. *Pelican in the Wilderness: a Naturalist's Odyssey in North America.* George Allen & Unwin, London.

Kerouac, J. 1957. *On the Road.* Viking Press, New York.

Leopold, A. 1949. *A Sand County Almanac*. Oxford University Press, New York.

Muir, J. 1916. *A Thousand-Mile Walk to the Gulf*. Houghton Mifflin, Boston.

Peterson, R. T. and Fisher, J. 1955. *Wild America: the Legendary Story of Two Great Naturalists on the Road*. Houghton Mifflin, Boston.

Sharkey, R. 1997. *The Blue Meteor: the Tragic Story of the Passenger Pigeon*. Little Traverse Historical Society, Petoskey.

Steinbeck, J. 1962. *Travels with Charley in Search of America*. Viking Press, New York.

Thompson, H. S. 1971. *Fear and Loathing in Las Vegas: a Savage Journey to the Heart of the American Dream*. Random House, New York.

Weidensaul, S. 2005. *Return to Wild America: a Yearlong Search for the Continent's Natural Soul*. North Point Books, New York.

CHAPTER 5

Atwater, C. 1838. *A History of the State of Ohio, Natural and Civil*. Glezen & Shepard, Cincinnati.

Baskett, T. S. (ed.) 1993. *Ecology and Management of the Mourning Dove*. Stackpole Books, Harrisburg.

Blockstein, D. E. and Tordoff, H. B. 1985. Gone forever: a contemporary look at the extinction of the Passenger Pigeon. *American Birds* **39**, 845–851.

Bucher, E. H. 1992 The causes of extinction of the Passenger Pigeon. *Current Ornithology* **9**, 1–36.

Clayton, D. H. and Price, R. D. 1999. Taxonomy of New World Columbicola (Phthiraptera: Philopteridae) from the Columbiformes (Aves), with descriptions of five new species. *Annals of the Entomological Society of America* **92**, 675–685.

Coy, P. L. and Garshelis, D. L. 1992. Reconstructing reproductive histories of Black Bears from the incremental layering in dental cementum. *Canadian Journal of Zoology* **70**, 2150–2160.

Diamond, J. M. 1989. Overview of recent extinctions. In Western, D. and Pearl, M. *Conservation for the Twenty-First Century*, pp. 37–41. Oxford University Press, New York.

Diamond, S. J., Giles, R. H., Kirkpatrick, R. L. and Griffin, G. J. 2000. Hard mast production before and after the chestnut blight. *Southern Journal of Applied Forestry* **24**, 196–201.

Forbush, E. H. 1905. The decrease of certain birds in New England. *Auk* **22**, 25–31.

Forbush, E. H. 1927. *Birds of Massachusetts and Other New England States*, Volume 2. Norwood Press, Norwood, MA.

Goodrum, P. D., Reid, V. H. and Boyd, C. E. 1971. Acorn yields, characteristics, and management criteria of oaks for wildlife. *Journal of Wildlife Management* **35**, 520–532.

Gysel, LW 1971. A 10-year analysis of beech mast production and use in Michigan. *Journal of Wildlife Management* **35**, 516–519.

McGee, W. J. 1910. Notes on the Passenger Pigeon. *Science* **32**, 958–964.

McKinley, D. 1960. A history of the Passenger Pigeon in Missouri. *Auk* **77**, 399–420.

Martin E. T. 1914. What became of the pigeons? *Outing* **64**, 479–483.

Smith, K. G. and Scarlett, T. 1987. Mast production and winter populations of Red-headed Woodpeckers and Blue Jays. *Journal of Wildlife Management* **51**, 459–467.

Whitney, G. G. 1994. *From Coastal Wilderness to Fruited Plain: a History of Environmental Change in Temperate North America from 1500 to the Present*. Cambridge University Press, Cambridge.

Wilson, E. S. 1934. Personal recollections of the Passenger Pigeon. *Auk* **51**, 157–168.

Wilson, E. S. 1935. Additional notes on the Passenger Pigeon. *Auk* **52**, 412–413.

CHAPTER 6

Ambrose, S. 1996. *Crazy Horse and Custer: the Parallel Lives of Two American Warriors*. Anchor Books, New York.

Di Sylvestro, R. 2005. *In the Shadow of Wounded Knee: the Untold Final Chapter of the Indian Wars*. Walker & Co, New York.

Earley, L.S. 2004. *Looking for Longleaf: the Fall and Rise of an American Forest*. University of North Carolina Press, Chapel Hill.

Gill, R. E. Canevari, P. and Iversen, E. H. 1998. Eskimo Curlew (*Numenius borealis*). In: Poole, A. and Gill, F. (eds.), *The Birds of North America*, No. 347, pp. 1–28. Academy of Natural Sciences and American Ornithologists' Union, Philadelphia and Washington DC.

Hacker, L. M. 1924. Western land hunger and the war of 1812: a conjecture. *Mississippi Valley Historical Review* **10**, 365–395.

Hemingway, E. 1927. *The Nick Adams Stories*. Charles Scribner's Sons, New York.

Krech, S. 1999. *The Ecological Indian: Myth and History*. W. W. Norton, New York.

Lockwood, J. A. 2004. *Locust: the Devastating Rise and Mysterious Disappearance of the Insect that Shaped the American Frontier*. Basic Books, New York.

Madson, J. 1982. *Where the Sky Began: Land of the Tallgrass Prairie*. Houghton Mifflin, Boston.

Muir, J. 1901. *Our National Parks*. Houghton Mifflin, Boston.

Ward, G. C. 1996. *The West: an Illustrated History*. Weidenfeld & Nicholson, London.

CHAPTER 7

Blockstein, D. E. 1998. Lyme disease and the Passenger Pigeon? *Science* **279**, 1831.

Buchanan, M. L. and Hart, J. L. 2012. Canopy disturbance history of old-growth *Quercus alba* sites in the eastern United States: examination of long-term trends and broad-scale patterns. *Forest Ecology and Management* 267, 28–39.

Ellsworth, J. W. and McComb, B. C. 2003. Potential effects of Passenger Pigeon flocks on the structure and composition of pre-settlement forest of eastern North America. *Conservation Biology* **17**, 1548–1558.

Speech by Robert F. Kennedy at the University of Kansas, March 18, 1968. http://www.jfklibrary.org/Research/Research-Aids/Ready-Reference/RFK-Speeches/Remarks-of-Robert-F-Kennedy-at-the-University-of-Kansas-March-18–1968.aspx.

CHAPTER 8

Avery, M. 2012. *Fighting for Birds: 25 Years in Nature Conservation.* Pelagic Publishing, Exeter.

Breeding Birds Survey reports. http://www.bto.org/volunteer-surveys/bbs/bbs-publications/bbs-reports.

Browne, S.J. & Aebischer, N.J. 2005. Studies of West Palearctic birds: Turtle Dove *Streptopelia turtur. British Birds* **98**, 58–72.

Browne, S. J., Aebischer, N. J. & Crick, H. Q. P. 2005. Breeding ecology of Turtle Doves *Streptopelia turtur* in Britain during the period 1941 to 2000: an analysis of BTO nest record cards. *Bird Study* **52**, 1–9.

McCarthy, M. 2009. *Say Goodbye to the Cuckoo.* John Murray, London.

Monbiot, G. 2013. *Feral: Searching for Enchantment on the Frontiers of Rewilding.* Allen Lane, London.

State of Nature 2013. http://www.rspb.org.uk/ourwork/science/stateofnature.

Weed Research Organisation. 1986. *Weed Guide.* Schering Agriculture, Stapleford.

Interview with the NFU president in the Guardian newspaper 28 July, 2013. http://www.theguardian.com/environment/2013/jul/28/weather-heatwaves-climate-change-uk-farming?INTCMP=SRCH.

To learn more about Duncan Farrington's farm visit http://www.farrington-oils.co.uk.

To learn more about Nicholas Watts' farm visit http://www.vinehousefarm.co.uk.

To learn more about Rob Law's farm visit http://www.youtube.com/watch?v=A5ecxONqFVI.

To learn more about the 'Barker Boys' farm visit http://lodgefarmwesthorpe.blogspot.co.uk.

Acknowledgements

Carry Akroyd is a friend who lives down the road, and I am pleased that she produced the beautiful cover for this book.

Elizabeth Allen, the RSPB librarian, made me welcome and a cup of tea, and fetched a few volumes from the cellar for me.

Thomas Avery came up with the phrase 'when the West was won the Wild was lost' in a café in Cambridge.

Tim Birkhead met me for lunch and gave me good advice.

Hugh Brazier found more mistakes than I thought was possible while he edited this book, and I am very grateful to him – those that remain are my own. He also improved my English, which surprised me less. I am very grateful to him for his editing, and for more than forty years of friendship.

Clair Castle of the Balfour Library in the Zoology Department in Cambridge helped me find my way around.

Rosemary Cockerill was understanding about me heading off to the USA for five weeks and then tolerant of me hiding away and writing for the rest of the summer. She commented on every chapter.

Chris Cokinos got there first with his book on extinct American birds – my original idea was a book on a similar theme so I had to think again. When I met him, by chance, in Tucson, he said we should keep in touch and he was kind enough to read a draft of Chapter 4.

My friends at the Cornell Lab of Ornithology, particularly Jessie Barry and Chris Wood, but also John Fitzpatrick, Matt Young and others, made me feel very much at home.

Jenny Dunn and Tony Morris told me about Turtle Doves and showed me several nests on a hot day in Essex in July. Their skill and dedication inspired me.

Alison Enticknap tracked down the date of Martha Grier's

death and also gave me some information about my own relatives. This started a train of thought that was important in the writing of this book. She also commented very helpfully on Chapters 6 and 7.

Duncan Farrington is the type of farmer of whom we need more, and his Mellow Yellow mayonnaise is delicious.

Michael Federspiel of the Little Traverse Historical Society put me on the track of Ernest Hemingway when I visited Petoskey, and was very helpful in lots of other ways.

Rhys Green shares an interest in this subject, and he pointed me in the direction of some useful papers, acted as a sounding board and commented helpfully on Chapter 2.

Roderick Leslie told me that I'd have to learn about trees to write this book, and he was right – although I fear he will think that there is a long way still to go.

Dawn McCarthy introduced me to her husband, Brian, who just happens to be a professor at Ohio University in Athens, an expert on old-growth forest and vice-chair of the American Chestnut Foundation. Brian was kind enough to tell me about Dysart Woods.

Michael McCarthy, an excellent writer and excellent companion, and a true friend, has encouraged me through friendly criticism and sharing bottles of claret.

I am grateful to Jim Martin of Bloomsbury for his support in the writing of this book.

Bev Pozega, in Wyalusing State Park, Wisconsin, shares my enthusiasm for Passenger Pigeons and was great to talk to.

This book was partly fuelled by mid-afternoon ice-creams from Mr Whippy, who almost daily would stop outside my house in the hot summer of 2013, and with his loud, jangly music broke whatever concentration I had mustered. I had to buy a cornet each time to make the stop worthwhile, and those brief interruptions often started me off on a fresher vein of thought which benefited my book. I only ate the ice-creams for the sake of my writing.

Index